Fire Resistance in American Heavy Timber Construction

Jesse Heitz

Fire Resistance in American Heavy Timber Construction

History and Preservation

Springer

Jesse Heitz
Chaska, MN
USA

ISBN 978-3-319-32126-4 ISBN 978-3-319-32128-8 (eBook)
DOI 10.1007/978-3-319-32128-8

Library of Congress Control Number: 2016938527

Printed on acid-free paper

This Springer imprint is published by Springer Nature
The registered company is Springer International Publishing AG Switzerland

"To my mother, my mentor and my greatest teacher, whose unwavering support has made all of this possible."

Introduction

The United States has proven itself to be a nation marked by a plethora of architectural styles. One could make the argument that the United States is effectively void of its own hallmark style or type of building. However, over the centuries of the American built environment's history, heavy timber construction came to dominate the landscape, serving as a quintessential American building type.

To the undiscerning eye, heavy timber construction may very well look mundane, if not outright bland, almost cookie cutter in their appearance. These massive masonry structures with exceptionally robust solid-sawn timber frames undoubtedly bear a similar resemblance to one another, giving an almost awe-inspiring and intimidating presence that screams utility and industry. However, these structures feature not only detailed and widely differing nuances, but embody a unique evolution of an elderly practice.

This book will serve to chronicle the extraordinary rise, decline, and resurgence of an iconic American construction type. Furthermore, this book will provide an in-depth technical analysis and a thorough exploration of the building type itself, one that highlights the outstanding level of innovation and durability harnessed by heavy timber construction.

Introduction

Contents

Contents

List of Figures

Chapter 1
American Heavy Timber Construction

The term American heavy timber construction is one that on its face is fairly self-explanatory. While it is largely unknown in many portions of the world, it is a phrase that immediately resonates with large swaths of the American public and within numerous professional sectors in this country. However, for the fire service, heavy timber construction immediately evokes notions of monolithic and iconic structures of nearly unprecedented scale. These buildings are the physical embodiment of the American Industrial Revolution, the timber and masonry backbone of the United States' rise to the climactic status of economic juggernaut.

Heavy timber structures mark this country from Eastern Seaboard to Western Seaboard. While generally of massive proportions, these buildings come in a variety of sizes and house a multitude of different occupancy types. Heavy timber construction is descended from one of the oldest known construction methods, post-and-beam construction, a type that can still be found across the Western world.[1] Heavy timber found a home in the United States for a number of reasons, not the least of which being its inherent identification with historical practice and arguably centuries of proven fire-resistant performance. Additionally, in the modern industrialized world, it makes exceptional use of readily available resources, all while providing innovative commercial building solutions and great advances in allowable height and area.

However, at its very core, heavy timber construction is a fire service designation, one that separates it from the litany of American building types. While the American firefighter is typically intimately familiar with heavy timber construction, it is critical that a general audience understands not only the many technical facets of heavy timber construction but also clearly understands how it is differentiated from the other building type classifications.

[1] George Walter Born, *Preserving Paradise*: *The Architectural Heritage and History of the Florida Keys*, (Charleston, SC: History, 2006), p. 102

J. Heitz, *Fire Resistance in American Heavy Timber Construction*,
DOI 10.1007/978-3-319-32128-8_1

Buildings are classified according to the tenets of the International Building Code (IBC), formerly the Uniform Building Code (UBC), which governs structures "erected or to be erected, altered or extended in height" into five distinct categories defined by material test performances.[2] Buildings are therefore required to meet the minimum requirements of the building type based upon the tested fire-resistive capabilities of the primary building materials used. However, it is important to note that structures can exceed the minimum requirements and yet remain in their established construction type.[3]

Overall, there are five types of building construction in the United States signified with Roman numerals. Within each type, with the exception of heavy timber, there are subdivisions which correspond to a given building's fire-resistance rating. Such a classification is based upon three key factors, the building's intended occupancy, height, and building area.[4] Essentially, these subdivisions represent a more specific rating of a building's hourly fire-resistance performance. Typically, these building types are informally divided into two groups, noncombustible and combustible. Types I and II compose the noncombustible variety, while Types III, IV, and V represent the combustible types.[5] Additionally, the types are organized according to their fire-resistance ratings, with Type I being the most resistant and Type V possessing the lowest level of fire resistance. However, it is important to mention that Type IV, which represents heavy timber, is a bit of a misnomer within the parameters of such an organization, as its ratings equal or exceed the two types that fall before it.

The Type I (Figs. 1.1 and 1.2) family of structures are generally high-rise structures of 75 ft in height or more, although some exceptions do apply and it is not uncommon to find examples that are within the range of 35–55 ft. These buildings are built out of concrete and steel, with the steel members being coated with a fire-resistant material, often times entombed in concrete. In the state of Minnesota, which adheres to the regulations established by the IBC, Type I-A buildings are classified as fire-resistive noncombustible construction and are often high-rise office buildings, hospitals, and detention centers. They represent the most stringent of fire performance rating requirements to protect such high-risk occupancies. The hallmarks of these buildings are ratings of 3 h for exterior walls and the structural frame, 2 h for floor and ceiling assemblies, and 1.5 h for roof assemblies.[6]

Type I-B is also listed as fire-resistive noncombustible construction and is often represented by mid-rise office buildings. These structures, while still of incredibly stout construction and featuring the same general design features as their more robust siblings, are held to a slightly lesser performance standard. Type I-B structures

[2] Technical Services Information Bureau, *IBC Building Types*, (Orange: Technical Services Information Bureau 2008).

[3] Technical Services Information Bureau, 2008.

[4] Technical Services Information Bureau, 2008.

[5] Technical Services Information Bureau, 2008.

[6] State of Minnesota. Department of Public Safety, *Construction Types—Definitions*, (St. Paul: State of Minnesota, 2015).

Fig. 1.1 The Carver County Jail in Chaska, MN, of Type I construction (Photograph courtesy of the Chaska Herald)

Fig. 1.2 Type I dry storage warehouse at Pearl Harbor (Photograph courtesy of the Library of Congress)

Fig. 1.3 Type II dating to the World War II era (Photograph courtesy of the Library of Congress)

feature ratings as follows: 2 h for the exterior walls, structural frame, and ceiling and floor separations, with 1 h for the ceiling and roof assembly.[7]

Type II buildings (Figs. 1.3 and 1.4) are also divided into two subcategories, Type II-A and Type II-B. Type II-A is termed as protected noncombustible construction and features fire ratings of 1 h for exterior walls, the structural frame, and floor, ceiling, and roof protection. Its less rigid counterpart, Type II-B, also known as unprotected combustible construction, also features construction out of noncombustible elements. However, the type possesses no added fire-resistant measures such as protective coatings.[8]

Type III buildings (Fig. 1.5), like many on the list, possess two subdivisions as well, Type III-A and III-B. Type III-A, protected combustible or informally known as "ordinary" construction, features brick or concrete block walls as well as floor and ceiling assemblies that are often protected by 1-hour fire-resistance-rated drywall. These buildings feature fire ratings of 2 h for exterior walls, 1 h for the structural frame, and 1 h for floor, ceiling, and roof protection. Type III-B, or unprotected combustible construction, while the older and more traditional form of Type III construction, harnesses virtually all of the same structural features, except that their

[7] State of Minnesota, 2015.

[8] State of Minnesota, 2015.

Fig. 1.4 Interior of a Type II warehouse building dating to the World War II era (Photograph courtesy of the Library of Congress)

roof and floor assemblies are unprotected. As such, Type III-B maintains a 2-h fire protection rating for exterior walls; their internal structural elements are not rated.[9]

The focal point of this book rests on Type IV construction (Figs. 1.6 and 1.7), better known as "heavy timber" or informally known as "mill construction." Unlike all of the other construction types classified by the IBC, heavy timber has no subdivisions. While heavy timber is a combustible construction type, its performance standards under fire conditions far exceed those of its close cousin, Type III-B construction. Its ratings are limited to an examination of its exterior walls and structural frame, which are rated to 2 h and 1 h, respectively.[10]

The final classification according to the IBC is Type V construction (Fig. 1.8). Again, this type features two subdivisions, Type V-A and Type V-B. Type V-A, or protected wood frame construction, is a combustible type often used in newer apartment buildings; it may resemble Type III in appearance only. Its lightweight wooden structural frame is protected, achieving fire ratings of 1 h for the exterior walls, the structural frame, and the floor, ceiling, and roof assemblies. Type V-B is simply

[9] State of Minnesota, 2015.

[10] State of Minnesota, 2015.

Fig. 1.5 A typical Type III residential building dating to the World War II era (Photograph courtesy of the Library of Congress)

known as unprotected wood frame construction. For these structures, their wooden members are exposed, effectively offering no fire resistance whatsoever.[11]

While heavy timber construction often bears a similar resemblance to its cousin, Type III-B construction, its ability to withstand the destructive effects of fire is well-documented. Heavy timber has been used well over a century in the United States to construct massive fire-safe buildings. Heavy timber buildings take many forms in the United States; while most are found serving as factories and warehouses,[12] heavy timber buildings also take the form of churches, breweries, and office buildings, among many others.

However, this building type, as the "mill construction" name implies, finds its genesis in its original occupancy classification. As such, these gargantuan structures were erected for the heavy-duty wear and tear necessary of industrial buildings. Its thick exterior and interior walls were composed of numerous courses of brick, and less often stone. Their interior structural frames were constructed of massive timber beams and posts, with incredibly thick planks for flooring.[13] Everything about a traditional heavy timber building screams robustly built, to the point of being structurally redundant (Fig. 1.9).

[11] State of Minnesota, 2015.

[12] Vincent Dunn, *The Strategy of Firefighting*, (Tulsa, OK: PenWell, 2007), p. 128.

[13] RLGA Technical Services, *Building Classification—Part 2: Construction Types*, (Arizona: Ronald Geren, 2006).

Fig. 1.6 Guardian Angels Church in Chaska, MN, a Type IV church. Originally built in 1885, but destroyed by fire in 1902. Pictured is the 1902 rebuild (Photograph taken by author)

The multi-foot-thick, often 2–3 ft, brick or stone walls of these buildings usually followed a tapered design in which the walls thinned as their height rose (Fig. 1.10). The stepped system of wall construction allowed for a better progressive transfer of the load from the top floors down to the bottom level of the building.[14] The impressive walls of heavy timber buildings were almost exclusively load bearing, and unlike other construction types, heavy timber structures seldom featured curtain

[14] *Principles of Building Construction: Combustible*, Tech. no. FEMA/USAF/NFA-PBC:C-SM. 2nd ed., National Fire Academy.

Fig. 1.7 Interior of a Type IV heavy timber Ford plant as viewed in 1966 (Photograph courtesy of the Library of Congress)

walls. While not necessarily common, lintels and arches were sometimes utilized to help transfer unusually heavy loads. Additionally, it is not uncommon to find these buildings with features such as bracing or support cables which further stabilized the stressed exterior walls.[15]

Interior walls, which are almost always load bearing, are usually composed of masonry and are continuously supported throughout the building. They are generally extended through each and every floor and contain pockets or grooves which accept the ends of beams, or joists of floor or roof assemblies.[16] Floors in these buildings are constructed out of multiple layers of thick planks, and often the top layer is a finished floor laid diagonally across the planks. Roof construction (Fig. 1.11), while still stout, is quite comparable to that of Type III, although they often have permeations for ventilation systems.[17]

To be sure, this may seem like a bit of an unusual idea, the notion that a building composed of combustible elements could offer inherent fire-resistance performance. However, that is exactly what heavy timber construction delivers, a proven system of structural design that, when built to code, resists fire exceptionally well. It has been found over many decades of study, as well as through trial and error, that when exposed to high heat, large-dimensional timbers char on the surface (Fig. 1.12),

---

[15] *Principles of Building Construction*: *Combustible.*

[16] *Principles of Building Construction*: *Combustible.*

[17] *Principles of Building Construction*: *Combustible.*

Fig. 1.8 Typical American lightweight residential construction commonly associated with Type V (Photograph courtesy of NIOSH)

which insulates and protects the unburned portion of the member.[18] Timber of significant size retains its mechanical integrity when exposed to heat; it is a poor conductor of heat, and unlike steel or iron, it does not expand or quickly lose its structural rigidity under the duress caused by heat.[19]

When it comes to heavy timber construction, one thing is certain—they are beyond strongly built. Aside from being rigorously built out of substantial materials, certain design features were employed to ensure that these masonry and timber giants could survive catastrophe. The large timber structural members were arranged

[18] Paul C. Gilham, *Fire-Resistive Design of Exposed Timber Structures*, (Tualatin: Western Wood Structures, 2015).

[19] Gilham, 2015.

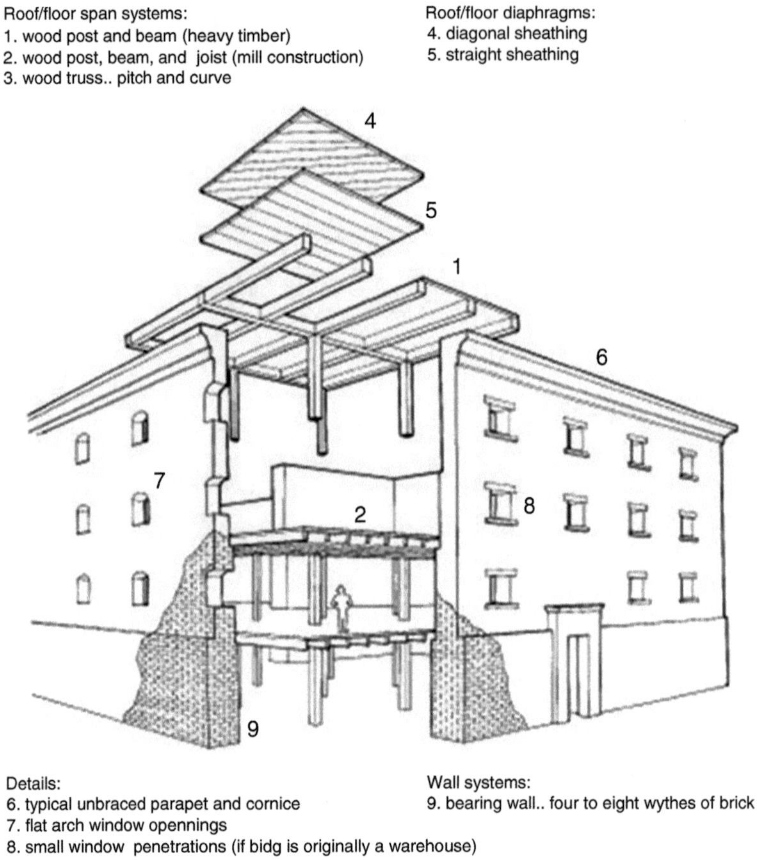

Fig. 1.9 An artist's rendering of a typical heavy timber structure's constituent components (Photograph courtesy of the CDC)

with the intention of eliminating sharp and protruding edges and void spaces. Additionally, designers sought to eliminate the potential of vertical fire spread through constructing floor assemblies a quarter of a foot thick.[20] When taken together, these unique and virtually unprecedented design features minimized points of ignition, prevented the passage of heat and vapor through the structure, and denied fire the opportunity to travel rapidly and undetected through plenums within the building.[21]

However, as with any structure, its use, and at times abuse, can wear down even these industrial castles. For example, heavy timber buildings used in the manufacturing trades often have flooring soaked in oil or other solvents, and the components and materials being processed themselves may be exceptionally flammable, creating a massive fire load. Add to that the era in which these goliaths were erected, and

[20] *Principles of Building Construction: Combustible.*

[21] Gilham, 2015.

Fig. 1.10 Interior walls of Guardian Angels Church in Chaska, MN (Photograph taken by author)

Fig. 1.11 Heavy timber roof assembly of Guardian Angels Church (Photograph taken by author)

slight deviations from code-based designs are not uncommon. Unfortunately, even minor deviations can undermine the integrity of the entire building. Building construction expert and 42-year veteran of the Fire Department of the City of New York, Vincent Dunn once commented that a true heavy timber building is quite rare and

Fig. 1.12 Charred heavy timber beam surviving from 1902 fire in Guardian Angels Church and reused in the rebuild (Photograph taken by author)

commenting specifically on floor assemblies stated, "This last requirement has escaped me. I have never see a factory or mill building with this floor. In fact, the floors of the heavy timber buildings I have observed looked pretty bad. Nevertheless, as the saying goes, if it walks like a duck and quacks like a duck, it is a duck."[22]

The currently embraced building codes identify what the ideal heavy timber construction building should look like. The type itself is to feature exterior walls that are of noncombustible materials, and the building materials on the interior of the structure are to be of solid or laminated wood. Minimum sawn nominal dimensions are required, and the presence of void spaces is expressly prohibited.[23]

More specifically, wood columns are to be either sawn or glued-laminated and are not to be less than 8 in. nominally in depth and 8 in. nominally in width when supporting floor loads. These members may be scaled back to 6 in. in nominal depth and 8 in. in nominal width when only supporting roof and ceiling loads.[24] All columns are to be continuous within the building or superimposed or connected in an approved fashion.[25]

Floor framing in a heavy timber building requires that beams and girders are of sawn or glued-laminated timber. Such members are required to be not less than 6 in.

---

[22] Vincent Dunn, *The Strategy of Firefighting*, (Tulsa, OK: Penn Well, 2007), p. 147.

[23] State of New Jersey, Division of Codes and Standards, *Types of Building Construction—International Building Code*, (Trenton: State of New Jersey, 2006), p. 77.

[24] United States Fire Administration, National Fire Academy, *Building Construction: Heavy Timber Construction* (*Type IV*), (Emmitsburg: National Fire Academy, 2013).

[25] State of New Jersey, 2006, p. 77.

nominal in width and 10 in. nominal in depth.[26] Any arches constituting the floor assembly and assisting in carrying the floor load that are either sawn or glued-laminated timber are mandated to be not less than 8 in. nominal in width and 8 in. nominal in depth. The floors themselves are to be without any void spaces and are to be composed of either sawn or glued-laminated planks, either splined or tongue and groove, of not less than 3 in. nominal in thickness, with 1-in. nominal dimension tongue-and-groove flooring laid atop the planks either diagonally or crosswise. Such a top flooring layer may be substituted for 15/32-in. wood structural panel or 1/2-in. particleboard or planks not less than 4 in. nominal in width. The flooring is to be laid so that no continuous joints are present except where supports exist. Additionally, floors are not to extend any closer than half an inch to the interior surface of the walls as to allow the natural expansion and contraction of the floor assemblies.[27] However, wood molding or corbelling of the masonry walls themselves is allowed to cover this gap.[28]

Roof framing, either wood frame or glued-laminated arches intended for roof construction and are not intended to carry floor loads, is to have timber members measuring 6 in. nominal in width and not less than 8 in. nominal in depth for the lower half. For the upper half, said timber members are to measure at least 6 in. nominally in both width and depth.[29] Otherwise, framed or glued-laminated arches springing from the roof or top of the walls, or framed timber trusses or any other roof framing element not supporting a floor load, are to measure at least 4 in. nominal in width and 6 in. nominal in depth. Spaced members are allowed to be composed of more than one piece, but must measure at least 3 in. nominal in thickness where blocked solidly throughout their intervening spaces or where such a space is tightly blocked by a continuous wooden plate that measures at least 2 in. nominal in thickness and is fastened to the members themselves. All other timbers composing a roof assembly must be at least 3 in. nominal in width, if a sprinkler system exists beneath the roof decking to protect such small members.

Roofs themselves are to be constructed without void spaces; as such any sort of drop ceiling or plenum is strictly forbidden. Roof decking is to be composed of sawn or glued-laminated planks that are either splined, tongue and groove, or a wood structural panel measuring at least 2 in. nominal in thickness or 1 1/8 in. nominal in thickness. Planks are required to be of at least 3 in. nominal in width, set on edge and laid according to the method prescribed for floors.[30] Currently, other forms of decking are allowed if they meet the same fire-resistance performance standards as the aforementioned variety.[31]

[26] United States Fire Administration.

[27] State of New Jersey, 2006, p. 77.

[28] State of New Jersey, 2006, p. 77.

[29] United States Fire Administration.

[30] United States Fire Administration.

[31] State of New Jersey, 2006, pp. 77–78.

Partitions are to be of solid wood construction and formed by no less than two layers of 1-in. matched boards or laminated construction that is at least 4 in. thick or of at least 1-h fire-resistance-rated material.[32] For exterior structural members where a horizontal separation of 20 ft or more exists, wood columns and arches adhering to the size specifications already prescribed by the code governing heavy timber construction are permitted.[33]

Heavy timber construction, despite its seemingly archaic origins, is a marvelous building type that showcases outstanding engineering principle and design, as well as upper echelon craftsmanship. When built within the aforementioned parameters, heavy timber is as sturdy and as fire resistant as a building featuring inherently flammable materials can be. Its evolution, much less its survival into the present day, is a fascinating tale.

[32] United States Fire Administration.

[33] State of New Jersey, 2006, p. 78.

Chapter 2
The Origins of Heavy Timber Construction

While it has been clearly established what heavy timber construction is, that is to say its design features and structural requirements have been transcribed, it is pertinent to explain how and why the construction type came into being. The origins of heavy timber construction are a multi-dimensional tale, an extraordinary saga that parallels the very history of the United States itself. The construction type truly manifested itself as the physical embodiment of the nation's rise from a veritable small-scale post-colonial status to undisputable industrial juggernaut.

Within a few short decades, the United States catapulted from relative obscurity on the global scale to a great nation marked by industrial ingenuity and enterprise,[1] and at the forefront of this rapid evolution was heavy timber construction. In order to fully appreciate why heavy timber construction rose to prominence, it is necessary to chronicle the development of the nation itself in the mid-nineteenth century. Economic development in the United States, or its Industrial Revolution, did not take place largely until just after 1820.[2] The great wave of industrial growth emanated from the great East Coast urban centers of New York, Boston, Philadelphia, and Baltimore. Within two decades, each of these cities possessed nearby industrial satellite cities. New York City had Paterson, New Jersey. Boston had Lowell, Massachusetts. Philadelphia had Reading, Pennsylvania, and Baltimore had Wilmington, Delaware.[3]

It was in these urban centers where large swaths of the nation's population resided. They were localities that flowed with capital and markets hungry for goods. By 1840, these eastern cities controlled approximately one-quarter of the United States' total manufacturing. From 1828 to 1860, the United States saw its GDP

[1] 'The State of the Manufacturing Trades.' *Scientific American*, XXV.13 (1871), p. 200.

[2] David R. Meyer, 'The Roots of American Industrialization, 1790–1860.' *Economic History Services*, (2003).

[3] Meyer, 2003.

J. Heitz, *Fire Resistance in American Heavy Timber Construction*,
DOI 10.1007/978-3-319-32128-8_2

grow by an annual average of 4.6 %.[4] During this period, one of the primary mechanisms for the profound expansion of the United States economy took place, the establishment of an extensive rail network.[5] Quickly, capital flowed into the rail lines from a variety of sources based out of these urban centers. In fact, between 1830 and 1839, investment in the nation's rail network increased at an average annual rate in excess of 31 %.[6]

The establishment of a modernized and efficient transportation network in the form of railroads, which were not dependent upon waterways or animal power, allowed for goods produced in the nation's manufacturing centers to reach distant markets. In short order, the East Coast manufacturing centers were able to capitalize on the sudden increase in the size of markets and the resultant increase in demand. Within a few years, industry exploded, with one Lowell-based company, Merrimack Manufacturing Company, which was an 1823 upstart, posting average annual profits of 24 % during the period from 1825 to 1845.[7] From 1805 to 1811, the aforementioned urban centers largely collected in the New York and southern New England area exceeded the national per capita patent filings by a threefold margin.[8] By 1860, this geographic area was producing over half of the nation's patents while possessing only one-fifth of the nation's population.[9]

To physically house this type of uncharted industrial and economic growth, suitable buildings were required. Early American industrial buildings were not entirely dissimilar from those in Europe. However, American mills and industrial buildings featured one stark contrast; instead of exterior walls filled with brick, stone, or wattle and daub, they were instead clad with clapboards.[10] In the United States, the great abundance of wood made it far more economical, if not more risky, to build in timber. However, this relatively lightweight construction method, and the exceptionally combustible nature of early manufacturing itself, led to fires being commonplace.[11] In the early-nineteenth century, post-and-beam construction, a precursor to heavy timber, emerged as a common building form for industrial buildings, a form that would better combat fire. This old world construction type featured masonry exterior walls, heavy timber beams and columns, and often 2-in.-thick timber floors sometimes covered in a 1-in. layer of mortar.[12]

---

[4] Rui M. Pereira and William J. Hausman, *Railroads and Economic Growth in the Antebellum United States*, (Williamsburg: College of William and Mary, 2014), p. 5.

[5] Pereira, p. 5.

[6] Pereira, p. 5.

[7] Meyer Weinberg, 'Rise of the Capitalist Class, 1790–1865.' *New History*, (2002).

[8] Kenneth L. Sokoloff, 'Invention, Innovation, and Manufacturing Productivity Growth in the Antebellum Northeast.' *American Economic Growth and Standards of Living before the Civil War*, (Chicago: University of Chicago, 1992), p. 352.

[9] Sokoloff, p. 352.

[10] Randolph Langenbach, *Better than Steel*? (*Part 2*): *Tall Wooden Factories and the Invention of "Slow-Burning" Heavy Timber Construction*, (Oakland: Conservation tech Consulting, 2010).

[11] Langenbach, 2010.

[12] Langenbach, 2010.

In 1835, an American building revolution took place in the form of the standardization of heavy timber construction. Leading up to that historic year, a number of fires had utterly decimated New England's textile mills, with flames routinely preying on the highly combustible contents and relatively lightweight construction of said structures. So commonplace were the destructive blazes that insurance companies refused to insure mills and factories, with the Monmouth Mutual Fire Insurance Company flatly stating, "Mills and Factories are rejected altogether…."[13] In the wake of these circumstances, a group of Rhode Island mill owners formed an organization which later became known as the Associated Factory Mutual Fire Insurance Companies and changed American architecture forever.[14]

Their summit in 1835 led to a standardized design for mill construction. The recommended design harnessed both the available resources, but utilized elements known for their proven fire-resistant qualities and for permanence.[15] With the official launch of the heavy timber construction type, it did not take long for the design to find widespread use in a variety of industries and occupancy types. Over the type's early years, the original recommended design was adjusted as necessary when issues arose that required a deviation in order to increase performance.[16]

By 1840, it appears that the heavy timber construction type had been truly standardized and streamlined. That year, the superintendent of mills in York, Maine, James Montgomery, commented on New England's mills, stating:

> Though the Mills in this country are not so high as those in Great Britain, they are generally very strong and durable. Instead of joists for supporting the floors, there are large beams about 14 inches by 12, extending across from side to side, having each end fastened to the side wall by a bolt and wall plate: these beams are about five feet apart, and supported in the centre by wooden pillars, with a double floor above. The under floor consists of planks three inches thick; the upper floor of one inch board. Some have the planks dressed on the underside, others have them lathed and plastered: the floor being in all four inches thick, is very strong and stiff. The average thickness of the side walls may be from twenty to twenty-four inches, and they are generally built of bricks. There are very few stone walls, free stone being scarce in this country.[17]

As heavy timber proliferated across the upper Eastern Seaboard, alternative fire-resistant building technologies were well known, but appear to have been largely avoided. As early as the 1850s, discussion about the use of iron for structural purposes had become common; however, there were great fears that cast-iron columns would fail under severe strains,[18] and when used in industrial occupancies with heavy fire loads and subsequently subjected to the intense heat of such a fire, the

---

[13] 'Monmouth Mutual Fire Insurance Company.' *Maine Farmer and Journal of the Useful Arts*, 7.51 (1839), p. 403.

[14] C. E. Paul, *Heavy Timber Mill Construction Buildings*, (Chicago: Engineering Bureau, National Lumber Manufacturers Assoc., 1916), p. 6.

[15] Paul, p. 6.

[16] Paul, p. 6.

[17] Langenbach, 2010.

[18] 'Fire-Proof Buildings.' *Circular*, 3.116 (1854), p. 462.

iron members would rapidly fail.[19] Moreover, it was common knowledge that hardwoods, such as oak, provided excellent strength and flexibility.[20] As such, an argument can be made that the adherence to heavy timber construction, even in the face of other building systems, could be due to both the widespread availability of suitable timbers[21] and heavy timber's proven performance under fire conditions.

Soon after its inception, heavy timber construction began to dominate New England cities as it slowly swept down the Eastern Seaboard. Providence, Rhode Island, the effective birthplace of the construction type, was one of the great pioneering cities that erected heavy timber buildings to bear the weight of its industrial growth. Coinciding with heavy timber's rise to dominance, Providence had its population triple between 1830 and 1865 and later doubling between the years of 1865 and 1880.[22] Providence's industrial growth initially started in 1811 when cotton was the primary trade; however, over the next few decades, the city's industry had diversified to the point it was labeled by one observer stating, "Providence manufactures everything from a carpet-tack to a locomotive."[23] Perhaps there was no better structural example of Providence's diversified nature than the Heaton and Cowing Mill (Fig. 2.1). This milling complex originally built in 1832, with additions through the 1980s, features a wide array of successive building design systems, including heavy timber.[24]

In Massachusetts, two great industrial centers, Woonsocket and Brockton, stand as excellent exhibitors of heavy timber construction. From 1835 to 1867, Woonsocket populated itself with numerous heavy timber textile mills, with a preference for stone over brick for the load-bearing walls. Most of these mills were relatively small, on average ranging from two to four stories tall and often of long and narrow rectangular shapes in order to capture the maximum amount of natural light.[25] In Brockton, the great rise in the construction of heavy timber structures did not largely take place until the 1850s and 1860s, when a great increase in the city's immigrant population fed a local economy primed to launch. However, the great surviving heavy timber buildings dated after 1870 (Fig. 2.2) are common in the city and feature a typically utilitarian appearance with broad gabled roofs and ranging in height from three to five stories.[26]

---

[19] 'Fire-Proof Structures.' *Flag of Our Union*, 12.26 (1857), p. 205.

[20] 'The Builder's Manual.' *Mechanics' Magazine, and Journal of the Mechanics' Institute*, 1837, p. 198.

[21] Winston Wallace Clement, *Standardization in the Lumber Industry: Trade Journals, Builder's Guides and the American Home*, (Philadelphia: University of Pennsylvania, 2014), p. 19.

[22] Richard E. Greenwood, 'Providence One Hundred Years Ago—The Industrial Heyday.' *City of Providence City Archives*, (2011).

[23] Greenwood, 2011.

[24] Edward Connors, *National Register of Historic Places Registration Form—Heaton & Cowing Mill*, (Riverside: Edward Connors and Associates, 2012), pp. 3–5.

[25] 'History of Fire and Fire Protection in Nineteenth Century Woonsocket.' *City of Woonsocket*, Woonsocket Fire Department.

[26] *MHC Reconnaissance Survey Town Report Brockton*, (Boston: Massachusetts Historical Commission, 1981), pp. 1, 14

Fig. 2.1 Heaton and Cowing Mill in Providence, Rhode Island (Photograph courtesy of Ned Connors)

Fig. 2.2 The W. L. Douglas Shoe Company in Brockton, Massachusetts, in the 1940s (Photograph by Jack Delano, courtesy of the Library of Congress)

As the nineteenth century progressed, so too did heavy timber's gradual migration southward. One of the great pockets of East Coast heavy timber construction was Paterson, New Jersey. Here, most of the mill and industrial heavy timber buildings were of heavy timber construction expressly built for utility rather than elegance. These buildings are considerably more plain than those existing in other localities. Paterson embodies the region's adherence to the construction type and a proven belief in the type's performance.[27] In 1871, a fire ripped through the city's Franklin Mill. Initial news reports stated that the fire had ostensibly been burning for quite some time before it was discovered, allowing for the development of quite an inferno. However, the blaze was kept to one wing of the mill, and its nearby exposure, the Hope Mill, was able to withstand the radiant heat.[28]

Toward the middle of the nineteenth century, heavy timber construction made its way into the American South. The patterns of heavy timber proliferation throughout the South tend to, like the North, follow its general economic development timeline. The South as a whole did not really start dramatically investing in its economy, that is to say by far its most dominant industry of cotton, until about 1840. In the short window between 1840 and 1860, the South had doubled its capital investment in cotton manufacturing.[29] Such a dramatic increase by virtue necessitated a resultant increase in infrastructure. To that end, heavy timber made its way through the American South.

To be certain, cotton has often caught the eye of many historians as the dominant industry of the American South, particularly the Antebellum South; however, in the second quarter of the nineteenth century, tobacco became a powerful industry in its own right, notably in Virginia[30] and North Carolina. Understandably, the American Civil War and resultant Reconstruction period placed a significant damper on building construction in general. However, in North Carolina, the city of Durham in particular, industry and new construction exploded in the 1870s. Cigarettes had become quite popular and the tobacco-producing regions such as Durham prospered.[31] A striking example of heavy timber construction's application in the South and its adaptation to tobacco production can be seen in Durham's American Tobacco Company Manufacturing Plant (Fig. 2.3). In this massive complex, which saw almost continual additions between 1874 and 1954, 30 buildings are heavy timber,

---

[27] Toni Ristau, *Mill Architecture in Paterson, NJ: A Culmination of the Empirical Tradition in Construction*, Proc. of Symposium on Industrial Archaeology, New Jersey, Paterson, (Paterson: Northeast Historical Archaeology, 1975), p. 59

[28] 'Great Fire at Paterson.' *The New York Times*, 14 December 1871, p. 1.

[29] Edward Pessen, 'How Different from Each Other Were the Antebellum North and South?' *The American Historical Review*, 85.5 (1980), p. 1125.

[30] Karen Lang-Kummer and A. Rebecca Harrison, *National Register of Historic Places Inventory-Nomination Form: Shockoe Valley and Tobacco Row Historic District*, (Richmond: Virginia Historic Landmarks Commission, 1982), p. 2.

[31] 'Cultivation of a Tobacco Empire.' *North Carolina Historic Sites*, North Carolina Department of Cultural Resources Office of Archives & History, (2014).

Fig. 2.3 The Bull Durham Tobacco Company's complex in Durham, North Carolina, in the mid-twentieth century (Photograph by Marion Post Wolcott, courtesy of the Library of Congress)

and while generally built in a simple utilitarian style, in later years newer buildings were adorned with decorative brickwork.[32]

While heavy timber construction generally seems to have followed a rather linear trajectory during the early years of its existence, it appears to have abandoned that trend by the middle of the nineteenth century. While it cannot be definitely ascertained as to why heavy timber construction's spread westward arguably progressed significantly faster than it did into the southern reaches of the United States, it can, however, be clearly seen that heavy timber was establishing itself in force in the Midwest, some 1,500 miles from the source, at around the same time it became popular in the South.

[32] Claudia R. Brown and M. Ruth Little, *National Register of Historic Places Inventory-Nomination Form: American Tobacco Company Manufacturing Plant*, (Raleigh: North Carolina State Historic Preservation Office, 2000), pp. 6–7.

Chapter 3
The Great Chicago Fire and Heavy Timber Construction

Chicago has long served as the unofficial gateway to the western half of the United States. This was particularly true during the second half of the nineteenth century. It was largely through this metropolitan nexus that commerce, and even technological innovation, flowed all the way to the Pacific Ocean. However, even in the current era in which the Western Seaboard has become densely populated and self-sufficient, the Upper Midwestern states, notably Minnesota, Iowa, Wisconsin, and Michigan, still look to the city of Chicago for direction.

While the level of dependence may very well have waned in the last half-century, during the American Industrial Revolution, Chicago's gravitas among its neighbors was undeniable, perhaps even inescapable. For a number of reasons that will be dissected in this chapter, Chicago's experience in building construction, and the related lessons stemming from the most destructive urban fire in American history, greatly influenced the built environment paths its neighbors in the Midwest elected to follow.

Chicago was first discovered by people of European extraction in 1673 by Jacques Marquette, a Jesuit missionary, and Louis Jolliet, a fur trader. The two explorers had been searching for the coveted Northwest Passage, but instead found a place that would become one of the largest and most influential cities in the United States.[1] This relatively barren plot of land located between the Chicago River and Des Plaines River, which flowed into Lake Michigan and the Mississippi River, respectively, ensured that whatever settlement took hold would become a regional hub due to its ready access to numerous significant waterways. Over two decades later in 1696, Jesuit missionaries established the Mission of the Guardian Angel with the hopes of converting the Native American populations, notably the Miami, Sauk, and Fox. For the next half-century, the location was a veritable hotspot for trappers and traders; however, by the mid-eighteenth century, the outpost had been largely surrendered to the Potawatomi tribe.[2]

---

[1] Owens, L. L. Owens, *The Great Chicago Fire*, (Edina: ABDO Pub., 2008), pp. 21–22.

[2] Owens, p. 22.

J. Heitz, *Fire Resistance in American Heavy Timber Construction*,
DOI 10.1007/978-3-319-32128-8_3

It was not until the 1770s that the first European-derived settlement was established on the site of modern Chicago by Jean Baptiste Point du Sable. By 1795, the United States, through the Treaty of Greenville, absorbed the surrounding area that constituted what in 1803 became Fort Dearborn, the United States westernmost military installation.[3] In 1816, the United States acquired part of the area from the occupying Native American tribes, and by 1818, Illinois became the 21st state. By the mid-1820s, as the United States began its westward expansion under the guise of Manifest Destiny, Chicago had become an outlying travel destination and bustling trading center.[4]

In 1833, Chicago was incorporated as a town with a total population of 350 residents. Within 4 years, Chicago's population had increased nearly 12-fold to 4170, and by 1837 it had been incorporated as a city.[5] Chicago by this point was well on its way to becoming the "Queen of the West." In the years that followed, the infrastructure, notably railroads, necessary to become a metropolis had been installed. By 1860, the city was prominent enough to host the Republican National Convention, which ultimately selected future president, Abraham Lincoln, as its nominee in the upcoming general election. In 1870, Chicago's official population bordered on 300,000, an unprecedented rate of growth for an American city.[6]

Chicago's meteoric rise to regional dominance and its explosive growth, which was fed by the influx of tens of thousands of immigrants each year, delivered by the ten railway lines that fed into the city, among other forms of transportation, created a distinct built environment, not seen elsewhere in comparably sized American cities.[7] In 1831, Chicago possessed a mere 12 houses.[8] The great opportunity and promise of the capitalist dream created by the incredible explosion of the city's growth also resulted in a boom of lightweight construction, ostensibly the only way to feed a rapidly growing city.

Chicago is often credited with establishing balloon frame construction. This method was a radical departure from the tried and true post-and-beam form of construction. Balloon frame construction saw the building of tenable structures requiring significantly less raw materials than traditional building methods. Balloon frame construction arranged a series of comparatively thin studs running vertically from basement to attic with simple weatherboarding fastened to the exterior face of the studs and lath and plaster applied to the interior face of the studs. This method effectively replaced the elaborate and time-consuming jointing of heavy timber frames with the easy application of nails. Effectively, the great rise of balloon frame construction fit the cheaper and faster modus operandi of a city such as Chicago which was growing at an exponential rate.[9]

---

[3] Owens, p. 24.

[4] Owens, p. 24.

[5] Owens, p. 24.

[6] Owens, p. 24.

[7] Owens, p. 25.

[8] Frank A. Randall, *History of the Development of Building Construction in Chicago*, (Champaign: University of Illinois, 1999), p. 8.

[9] Charles Waldheim, Katerina Rüedi, *Chicago Architecture: Histories, Revisions, Alternatives*, (Chicago: University of Chicago, 2005), p. 56.

Fig. 3.1 Typical 1880s balloon frame home showing extensive damage from unimpeded vertical fire spread (Photograph taken by the author)

Longtime Chicago resident, Charles Cleaver, recounted his observations of the building trades, and the rise of balloon frame construction (Fig. 3.1), in the city at the time of his arrival in 1833 by stating, "Heavy timber for frame buildings soon after that came into disuse, as it was found the present way of putting up buildings was much stronger and better. It was then called balloon framing. G.W. Snow, an old settler, had the credit of first originating the idea."[10]

Another longtime Chicagoan, William Bross, recounted Chicago's predominantly balloon frame identity at the time of his arrival in 1846, by stating:

> Stores and dwellings were, with few exceptions, built in the balloon fashion....Posts were placed in the ground at corners, and at proper distances between them blocks were laid down singly or in cobble house fashion. On these foundations were laid, and to these spiked, standing on end, 3″×4″ scantling. On these sheathboards were nailed, and weatherboards on the outside of them; and lath and plaster inside, with the roof completed the dwelling or store. This cheap, but for a new town, excellent, mode of building, it is claimed was first introduced, or if you please, invented in Chicago, and I believe the claim to be true.[11]

After 1845, what could be called Type III construction had been introduced into the city's central business district. In wall assemblies, weatherboard exteriors were replaced with brick; however, lightweight wood members were still utilized in floor and roof assemblies. Beginning in the 1850s, the use of Lemont limestone, informally known as "Athenian marble," came into fashion. However, its use was gener-

[10] Sally Ann McMurry, Annmarie Adams, *People*, *Power*, *Places*, (Knoxville: University of Tennessee, 2000), p. 14.

[11] McMurry, p. 14.

ally restricted to rather pretentious structures in the business district. For Chicago, the limited adoption of masonry exterior walls was an ode to fireproofing, a tenet critical to any industrial city.[12]

From approximately 1855 to 1870, Chicago took another leap forward in fire protection when it became popular to entomb the exterior of fairly extravagant structures with cast iron. The first structures to feature iron casing were located on the Lloyd Block at the intersection of North Wells Street and West Randolph Street. In 1856, a block of buildings featuring iron casings was built along East Lake Street.[13] In 1868, a protective coating for wooden floor assemblies was developed and tested in Chicago, and when compared with a floor assembly made of iron and brick, it was observed that the timber held up satisfactorily.[14]

While Chicago steadfastly held onto its balloon frame construction method despite the introduction of more advanced building construction methodologies, the Chicago timber trade continued to propel the city's lightweight building growth. The insatiable demand for houses, grain elevators, warehouses, shops, churches, and stockyards kept the saw mills quite busy producing small-dimension timbers. The city's dramatically swelling infrastructure also made ample use of lightweight timber. By 1871, Chicago possessed some 88 miles of paved streets, with 57 miles of that being paved with pine. Additionally, the city had a total of 561 miles of wooden sidewalks.[15] From rooftop to sidewalk, Chicago was a city of lightweight wood. Such dedication to haste over the construction of sound and proven building systems effectively set the stage for the visitation of catastrophe upon Chicago.

During the summer and spring of 1871, Chicago experienced dozens of serious fires. During those months, a number of periodicals commented on a veritable brewing storm of disaster that almost seemed inescapable. The *Journal of Commerce* commented:

> For nearly fifteen weeks there had not fallen enough rain to penetrate the earth one full inch…the city was heated, dry, and parched. Indeed, all through the West, fires were devastating extensive forests and destroying ripening crops, driving frontier settlers from their cabins and even overwhelming entire villages. For days, the prevailing atmosphere of our city seemed ready to kindle into a blaze.[16]

The *Chicago Tribune* also commented on the state of building construction in the city, almost criticizing the city's seeming preference for lightweight construction over the elements found in heavy timber construction, stating, "In a city where time was everything, and durability was not a matter much considered, street after street was lined with wooden buildings, not with oaken beams and floorings, but an aggregation of flimsily constructed and inflammable pine."[17]

---

[12] Randall, p. 8–9.

[13] Randall, p. 9.

[14] 'Test of Fire-Proof Flooring' *Railway Times*, 5 December 1868, p. 388.

[15] Owens, p. 26.

[16] Owens, p. 27.

[17] Owens, p. 27.

In the first week of October alone, the Chicago Fire Department responded to 20 structure fires.[18] Following up on a particularly dry season complete with the requisite conditions for the outbreak of fire, a Chicago-based correspondent for the *New York Times* reported:

> The wind was blowing very fresh and the flames spread with almost incredible rapidity, and in a few minutes the entire structure was a mass of fire. The immediate vicinity is built up mainly with small wooden tenement-houses and two-story frame buildings occupied as groceries, saloons. The inmates of many of these houses, startled from slumber, had barely time to rush out in their scanty night attire, leaving their household goods to destruction.
>
> In several instances children were hastily wrapped in blankets and quilts to break their fall, and thrown from the second-story window to the ground. When the alarm sounded for this fire, another of considerable magnitude was burning on Wells Street, near Adams. Several engines were necessarily kept at work upon this fire. The rest of the engines in the city were soon on the ground, but before they arrived the fire had spread over so large an area, and was so rapidly spreading, that efforts seemed of little avail....
>
> The scenes in the vicinity of the conflagration are indescribable. Half of the population of the city seems gathered there. The tugs in the river are engaged in towing to a place of safety the vessels moored in the neighborhood, while locomotives are hastily pulling out the great number of cars standing on the track in flames.[19]

The *Chicago Daily Tribune* ominously reported on the morning of 8 October of the aforementioned fire that burned several blocks the day before, stating:

> The sounding of the fire alarm from Box No. 248 at about 11 o'clock last night, was the solemn prelude to one of the most disastrous and imposing conflagrations which has ever visited a city which has already enrolled in her annals numbers of such visitations, many of them so terrible that they can serve as eras in her history. For days past, alarm has followed alarm, but the comparatively trifling losses have familiarized us to the pealing of the Court House bell, and we had forgotten that the absence of rain for three weeks had left everything so dry and inflammable a condition that a spark might start a fire which would sweep from end to end of the city.[20]

Half a day after the issuance of such prophetic words, Chicago's worst nightmare came true. Between 8:00 and 9:00 pm on the evening of 8 October, a fire started in the barn of the O'Leary's, whose property was located around 137 DeKoven Street. Initially, a guard keeping fire watch atop the courthouse noticed the fire and sounded the alarm, but misidentified the actual location of the fire as occurring near the intersection of Canalport Street and Halsted Street, instead of its true location at DeKoven Street and Jefferson Street. As the fire progressed, a nearby shopkeeper tried to sound a fire alarm, but the alarm box was new and nonfunctional.[21]

Taking advantage of the unusually dry conditions and Chicago's highly combustible construction, the fire quickly spread to neighboring properties. By 10:00 pm, seven local fire companies had responded to the scene; unfortunately, with nearly an hour and a half of burn time, the fire had become largely uncontrollable. Fanned by

---

[18] Owens, p. 27.

[19] 'The Fire Fiend.' *The New York Times*, 8 October 1871, p. 5.

[20] Owens, p. 28.

[21] Owens, p. 31.

strong winds from the Southwest, the fire quickly skipped across the Chicago River, following a trail of oil and debris scattered across the surface of the water. Soon the fire latched onto and proceeded to rip through the wooden warehouses and lumberyards that populated the riverbank. Before long, the fire reached the gasworks, which exploded around midnight, fueling the fire even further.[22]

On the afternoon of Monday, 9 October, the fire continued unabated. So fierce was the fire spread that Chicago's mayor, Roswell Mason, issued telegrams to both President Ulysses S. Grant and Lieutenant General Philip Sheridan, asking for immediate aid to both fight the fire and maintain the peace. By 2:00 pm that day, the fire jumped the Chicago River again, this time visiting devastation upon the city's North Side before advancing through the city's South Side. During the fire's run through the South Side, it ravaged the waterworks, effectively disabling the city's water supply and stymieing remaining firefighting efforts. So aggressive was the fire's spread and so great was the continued threat that upon General Sheridan's arrival that day, he ordered the demolition by explosion of large swaths of buildings along Wabash Avenue and Congress Street.[23] Despite such a drastic tactic, the demolition failed to halt the progress of the blaze. It wasn't until the skies opened up and delivered much-needed rain during the overnight hours that the fire was finally extinguished.[24]

After nearly a week and a half following the conclusion of one of the worst disasters in United States history, the *New York Times* printed a vivid description of the destruction. The piece stated:

> The Chicago ruins never fall upon the eye. For three days I have wandered among these wrecks of a great city, and they are still as new to me as they ever were. To those familiar with the city, it is enough to say briefly that the fire destroyed everything within an area bounded by certain streets, but to a stranger this statement conveys no information whatsoever.
>
> With what terrible completeness the flames did their work of destruction can only be realized by hours of inquiry among the ruins. This morning I crossed to the North Side by Kinzie-street Bridge, and skirted the southern edge of the district to the lake. The desolation is awful. The fire was here most intense, or fed upon more inflammable material, and everything is absolutely eaten up. There is no debris in the streets; there is none except a few scraps of iron or a few remnants of bricks upon the sites of the houses; the fire had gained such intensity here that it had literally devoured whole blocks. To the north could be seen only a waste of ashes, with here and there portions of calcined [sic] walls standing as monuments of the ruins. Reaching the street nearest the lake, and turning northward, it was still a wilderness of ruins behind me, before me, and upon my left. Far away to the northwest I could see the vast remnants of St. Michael's Church, and directly before me the scorched foliage of Lincoln Park. As I walked it was everywhere the same sad scene-nothing anywhere of what had been stately dwellings and splendid churches but a part of crumbling wall at the best. But even of these there were but few, as the fire had consumed all, and there was but heaps of ashes where once houses had stood.[25]

[22] Owens, p. 32.

[23] 'The Very Latest.' *The New York Times*, 10 October 1871, pp. 1.

[24] Owens, p. 39.

[25] 'The Chicago Calamity.' *The New York Times*, 21 October 1871, p. 11.

Fig. 3.2 The aftermath of the Great Chicago Fire (Photograph courtesy of the Library of Congress)

After over 2 days of raging, the fire had subsided, and had utterly devastated the city (Figs. 3.2 and 3.3), clearing a strip roughly 4 miles long and a mile wide. Reports from the field indicated that the fire had burned so intensely that one of the ravaged structures which contained stocks of steel and iron had become so heated that the stock of raw metal had fused together.[26] In its wake, the fire had killed 300 people; displaced 100,000 people, which was a third of the city's population; and had reduced approximately 18,000 buildings to either rubble or ashes. The fire destroyed some $200,000,000 worth of property, virtually one-third of the city's entire valuation.[27]

In the aftermath of such a catastrophe, significant analysis regarding the design and construction of fire-resistant structures took place. To be sure, significant devotion was paid to the causes that sparked such a blaze. Observers almost universally noted the unseasonably dry conditions, the appearance of a gale which fanned the flames, and the presence of large scores of lightweight wood, both in lumberyards and composing structures.[28] One observer even went so far as to forcefully argue that the fire was the result of coordinated arsonists.[29]

However, the one issue that cropped up among informed commentators was Chicago's lightweight building identity. One observer noted, "Our buildings were not constructed as substantially as they should have been. The few standing walls

[26] 'Chicago: Curious Effects of the Heat of the Conflagration.' *The New York Times*, 17 November 1871, p. 2.

[27] Jo Ann Rayfield, 'Tragedy in the Chicago Fire and Triumph in the Architectural Response.' *Northern Illinois University Library*. Northern Illinois University, (1997).

[28] 'Chicago: Progress in Clearing Away the Rubbish at the Burnt District.' *The New York Times*, 8 November 1871, p. 6.

[29] 'Chicago: Curious Effects of the Heat of the Conflagration.' *The New York Times*, 17 November 1871, p. 2.

Fig. 3.3 Photograph showing further damage from the Great Chicago Fire (Photograph courtesy of the Library of Congress)

seem to attest to this fact. But the pressure of the gale, intensified by the heat, was fearful. Once inside of a building the two forces raged so madly that no edifices could withstand them."[30] In the months that followed the fire, a near-architectural pandemonium erupted, in which current building methods were strenuously contested. One post-fire analysis commented that prior to the Great Fire, the city of Chicago contained approximately 60,000 lightweight pine structures with only a few thousand buildings constructed out of brick or stone.[31] One New York-based observer wrote:

[30] 'Chicago: Progress in Clearing Away the Rubbish at the Burnt District.' *The New York Times*, 8 November 1871, p. 6.

[31] 'Mayor Medill: His Inaugural Message to the New Council of Chicago.' *The New York Times*, 8 December 1871, p. 2.

> It is not worth anybody's while to tell us that fires in dwellings are comparatively infrequent, and that the destruction of handsome or costly private houses is so rare an event as to make insurance almost unnecessary. We understand all this, and do not raise any question upon such points. But, if nothing else is taught us all by the Chicago fire, we have learned that in the midst of our best built cities there lurks a possibility of widespread disaster, which conjunction of certain favoring circumstances may at any moment precipitate upon us. And if these elements of danger do exist, and are only waiting their opportunity to break loose; if a high wind, a cold night, a demoralized or exhausted fire department, a scarcity of water should; here form a partnership with the flames, does any one doubt that the result would be just such a mowing down of millions as Chicago has seen? It will not answer, therefore, to rest secure in the connection that, because, heretofore, the average method of construction has stood the test of moderate fires, there is no necessity of making adequate provisions against extraordinary combinations of other elements with fire. We want no repetition of the Western horror here.[32]

Once again, Chicago was embarking upon a population boom, one that saw continued exponential growth that resulted in a total population of 500,000 by 1880 and a total population of 1,000,000 by 1890.[33] While the city was just beginning its second phase of unprecedented growth, the overwhelming desire to rapidly rebuild a city in ruins manifested itself in a number of ways. Some large facilities found it more financially expedient to simply form their own full-service fire brigades including fully equipped fire engines,[34] instead of purchasing insurance or erecting more fire-resistant structural designs. On the other end of the spectrum, numerous calls for the outright prohibition of wood frame construction within city limits were fielded.[35]

This resulted in widespread calls for more robust construction within the city, particularly greater use of brick and iron.[36] Detailed analyses of wall construction were conducted and concluded that brick exterior walls were of an absolute necessity. Further arguments were made that timber should be removed from such assemblies in favor of iron, which it was argued possessed few of timber's drawbacks.[37] Calls for the adoption of iron girders with brick infill for the construction of floor assemblies became commonplace.[38] Effectively, Chicago had been primed for the use of masonry and iron as means of constructing its new brand of fire-resistant structures. As such, Chicago grasped onto its traditional guiding principle regarding building construction and efficiency in both cost and time.[39]

---

[32] 'The Secret of Fire-Proof Construction.' *The New York Times*, 26 November 1871, p. 4.

[33] Alice Sinkevitch, Laurie McGovern Petersen, *AIA Guide to Chicago*, (New York: Harcourt Brace, 1993), p. 5.

[34] 'Chicago: High Rates of Insurance—Bad Faith of Insurance Companies.' *The New York Times*, 11 November 1871, p. 4.

[35] 'The City of Chicago: Slow Progress of Rebuilding.' *The New York Times*, 8 December 1871, p. 2.

[36] 'Fire-Proof Cities.' *Every Saturday: A Journal of Choice Reading*, 4 November 1871, p. 434.

[37] 'The Risk of Buildings from Fire.' *Scientific American*, XXVII.11 (1872), p. 165.

[38] 'The Secret of Fire-Proof Construction.' *The New York Times*, 26 November 1871, p. 4.

[39] Sinkevitch, p. 5.

One month after much of the city had been rendered into ash, streets chalked full of brick- or stone-clad buildings were erected. Wabash Avenue, Dearborn Street, Madison Street, Water Street, Clark Street, Adams Street, Washington Street, and many others that bore the brunt of the devastation were quickly reborn in brick.[40] One year after the Great Fire, some $50 million worth of new buildings had been constructed in Chicago, largely of Type III construction. Furthermore, between 1872 and 1879, over 10,000 new construction permits had been issued, and by 1891, some $316,000,000 had been pushed into new building construction.[41]

In late November of 1871, one observer, almost annoyed with the new brick cookie-cutter appearance of downtown Chicago, wrote:

> So far as the progress of rebuilding has extended up to the present time, the architectural difference from Chicago as it was is strikingly noticeable. A general condition of enforced economy, together with a determination to secure massive permanency at the expense of elegant ornamentation, has brought about an almost uniform plainness. With few exceptions, the main walls are of the uncertain colored brick, peculiar to this locality, the fronts of red or white pressed brick, or of the painted or "pointed" style, the trimmings of stone or iron, and the cornices chiefly of brick. This excessively plain style, however, only obtains in the buildings now under way. The leading architects are perfectly swamped with plans and details for more elaborate and pretentious structures, to be commenced early in the Spring, so that the present is hardly the time to speculate on the appearance of Chicago when rebuilt. Two hundred and twelve permanent brick and stone buildings are in course of erection in the burnt district of the South Division, covering a total street frontage of 19,715 feet, or about three and a half miles.[42]

While Chicago almost wholeheartedly latched onto Type III construction, it also paid significant attention to alternative and ostensibly more robust fire-resistant construction methods. After all, in 1855 the city had erected the United States Post Office and Courthouse building, which featured heavy masonry walls and iron structural members; unfortunately as unprotected iron or steel is prone to do when exposed to intense heat, the iron beams and joists buckled, and the building collapsed.[43] To further complete the city's forceful turn toward the abundance and simplicity of metal as a means of fire protection, there were theories of protecting floors with layers of sheet iron covering the structural members of the floor assemblies.[44]

What is so interesting about the Great Chicago Fire, and what makes it such an exceptionally critical event with regard to the history of heavy timber construction, is the continued avoidance of the use of heavy timber construction within the city, a proven method of building fire-resistant structures, instead opting to pioneer new and arguably revolutionary methods of building construction. Chicago effectively looked to the Nixon Building as the model for a truly fire-resistant building.

---

[40] 'Chicago: Progress in Clearing Away the Rubbish at the Burnt District.' *The New York Times*, 8 November 1871, p. 6.

[41] Rayfield (1997).

[42] 'Rebuilding Chicago.' *The New York Times*, 29 November 1871, p. 2.

[43] Randall, p. 10.

[44] 'Fire-Proof Buildings—Views of the Sculptor, Powers.' *The New York Times*, 30 December 1871, p. 4

Fig. 3.4 A late-nineteenth-century photograph of the Nixon Building (Photograph courtesy of W. S. Wheeler)

The Nixon Building had been started in 1871, was located near the intersection of LaSalle Street and Monroe Street (Fig. 3.4) and was a five-story building designed for maximum fireproofing. The walls consisted of heavy masonry divided into deep and narrow piers spaced approximately 6 ft apart. The structural frame was made out of cast-iron columns and girders which held wrought-iron floor joists and roof rafters. Each bay spanned some 16 ft, and the upper sides of each floor were covered by 1 in. of concrete. To further limit vertical fire spread, ceilings were covered in an inch of plaster of Paris.[45] Nearing completion when the Great Fire swept through, the building survived the fire and was completed before the year ended.[46]

The Nixon Building, due to its reputation for being one of the very few buildings to have survived the destruction, and its nature which proved capable of harnessing

[45] Carl W. Condit, *The Chicago School of Architecture: A History of Commercial and Public Building in the Chicago Area, 1875–1925*, (Chicago: University of Chicago, 1966), p. 23.

[46] Randall, p. 12.

the latest in technology and a speed and ease of construction, became the archetype for future fire-resistant structures in Chicago, which began to multiply in the decade following the Great Fire. What is most curious is that for half a century, heavy timber construction had proven itself to be among the most fire-resistant construction types in existence within the United States. However, Chicago's general avoidance of heavy timber is strangely understandable as the construction type itself is costly in terms of building time, technical skill, and raw materials. One experienced heavy timber framer described the general process for preparing a single member for placement just after the turn of the twentieth century by stating:

> The first step in the process is to scaffold your timber so that it will lie straight and as nearly level as possible, and so that you and your men who follow may work over it in a comfortable position. That done, suppose we have a corner post to lay out which is 8 1/2 [inches] by 8 1/2 [inches] by 16 feet, and from shoulder to shoulder of tenons is 15 feet. I would select the two best faces that give nearest a straight corner, taking a corner that is hollow rather than one that is full. Then I set one square on across the best face, far enough from the end of a tenon, and measure 15 feet towards the other end, making an irregular mark across the face at this point with a heavy pencil as I did at the other end. I then set my second square on this mark and look over the squares. Just here comes in the nice point in unwinding timber. If at the first glance over the squares they should be very much in wind, then adjust the difference at each end by dividing. But this rule does not always work, for the wind may all be in the last two or three feet of the stick-more likely than not at the butt end. You will soon learn by looking over the faces of the timber to locate the cause or place of the wind. You will soon learn also that it requires but a slight change to adjust the squares so that there may be little cutting necessary in making the plumb spot. But to go on: with your adze or chisel level off across the face of the timber as much as you think will be necessary to bring the lines right in the end. While at this end of the timber spot on the side face, then go to the other end and unwind with the spot already completed. After making the plumb spot on the side face take your scratch awl and point with 2-inch face each way from your plumb spot, going around the four faces of the timber. Line through these points and work from the lines laying out.[47]

While this single process in the erection of a heavy timber building reinforces the notion that such a construction type is both time and resource consuming, two things which were simply unacceptable in Chicago's hustled building climate, it only adds to the reality that such complicated and laborious methods existed for a reason. While Chicago's industrialists, merchants, engineers, and architects all clamored for the latest and greatest fire-resistant designs, in essence mimicking the then iconic yet without fire load Nixon Building, which itself was a close descendant of the ruined Post Office and Courthouse building, they failed to recognize that out of the handful of surviving buildings, heavy timber was present.

The first structure was the Old Water Tower (Fig. 3.5). Built from 1867 to 1869, and designed by William Boyington, this castellated gothic-styled structure stood firmly in the path of the Great Fire. Constructed out of solid blocks of rough-faced yellow limestone, this still standing structure also featured a foundation of 168

[47] Fred T. Hodgson, *Light and Heavy Timber Framing Made Easy; Balloon Framing, Mixed Framing, Heavy Timber Framing, Houses, Factories, Bridges, Barns, Rinks, Timber-Roofs, and All Other Kinds of Timber Buildings*, (Chicago: F.J. Drake, 1909), pp. 167–169.

Fig. 3.5 Chicago's Old Water Tower (Photograph courtesy of the Library of Congress)

concrete-filled piles topped with twelve 12-in. timbers. The Old Water Tower, unlike its neighbors, survived the blaze intact, despite much postfire analysis that sharply criticized the vulnerability of timber members in masonry wall assemblies.[48] While not a true heavy timber building, the Old Water Tower's survival proved to be one piece of evidence suggesting that heavy timbers and masonry, when carefully designed and thoughtfully constructed, could perform as well as their masonry and iron counterparts.

However, it was a second structure, a true heavy timber building that stood at the epicenter of the fire and survived with little damage, that proved even a seemingly archaic building system such as heavy timber construction could still serve its

[48] City of Chicago, *Landmark Designation Report*, (Chicago: Commission on Chicago Historical and Architectural Landmarks, 1984).

Fig. 3.6 A view of Old St. Pat's exterior and interior during the mid-twentieth century (Photograph courtesy of the Library of Congress)

intended purpose. Located at 700 West Adams Street sits Old St. Patrick's Church (Fig. 3.6). Designed in 1852 by pioneering Chicago architect Augustus Bauer with the help of Asher Carter, this round-arched Romanesque Catholic church had its cornerstone laid in 1853[49] and, true to heavy timber form, was not dedicated until December of 1856. This large two-story brick and stone heavy timber church seated a congregation of approximately 1200 people[50] and was in full operation at the time of the fire and yet miraculously survived.[51]

While an argument cannot necessarily be made that heavy timber construction was superior to masonry and iron, the first incarnation of Type II construction, what can be clearly seen from the easily discernible available evidence is that heavy timber construction was able to survive one of the worst conflagrations in American history. The survival of a limited number of structures also demonstrates that the sphere of American architecture had long known of a proven fire-resistant construction type and that such an event further cemented heavy timber's place as a worthy building type that had not yet been relegated to the dust bin of architectural history. The question that remained was, as Chicago grew in clout by the day, would heavy timber be altogether replaced by more modern and efficient building types?

[49] Denis R. McNamara and James Morris, *Heavenly City*: *The Architectural Tradition of Catholic Chicago*, (Chicago: Liturgy Training Publications, 2005), p. 12.

[50] A. T. Andreas, *History of Chicago*: *From the Earliest Period to the Present Time*, (Chicago: A.T. Andreas, 1884), p. 294.

[51] Robert P. Ledermann, *State Street*: *One Brick at a Time*, (Charleston: History Press, 2011), p. 20.

Chapter 4
Heavy Timber in the Midwest

In the storied history of heavy timber construction, its proliferation across the Midwest is a particularly interesting case. The Midwest, while lacking the East Coast's population density and overall industrial might during the golden age of heavy timber, proved to be a region that wholeheartedly latched onto the construction type. The region also continued to build in heavy timber well into the era of concrete and steel. Two localities within the Midwest appear to have almost single-handedly propelled heavy timber's dominance in the region. These two localities are Minnesota's Twin Cities, notably Minneapolis, and St. Louis, Missouri.

The development of the Twin Cities into one of the Midwest's industrial juggernauts began rather early. Such an industrial evolution found its origins in the 1820s, when the United States government established both a sawmill and a grist mill along St. Anthony Falls to supply the nearby Fort Snelling. The milling industry would soon become the principal industry for Minnesota, due in no small part to the abundance of necessary natural resources.[1]

Despite what Minneapolis and St. Paul would later become, regional powerhouses, Minnesota's milling industry generally appears to have worked its way northward through the state toward Minneapolis (Fig. 4.1). In 1846, a fairly massive and robust four-story mill was constructed in the little town of Pickwick, just south of Winona, Minnesota. This particular mill, aptly named the Pickwick Mill, became one of the state's most prodigious flour mills, running 24 h a day during milling season. It later produced flour that fed Union troops during the American Civil War. The rapid development of the Minneapolis milling district in the 1870s and 1880s relegated the Pickwick Mill to clear second-rate status, but nonetheless, the Pickwick Mill was an interesting example of early Midwestern heavy timber construction and a physical foreshadow of the milling empire that was to come.[2]

---

[1] Scott F. Anfinson, 'Archaeology at the Riverfront: Unearthing the Invisible.' *Minnesota Historical Society*, (2003), p. 322.

[2] 'Solid Geometry: The Pickwick Mill.' *Preservation Matters*, 8 (1992), pp. 1, 7.

J. Heitz, *Fire Resistance in American Heavy Timber Construction*,
DOI 10.1007/978-3-319-32128-8_4

Fig. 4.1 Schech's Mill located near La Crescent, Minnesota (Photograph courtesy of the Library of Congress)

Following along southern Minnesota's fertile corridor to Minneapolis was the Cannon River Valley. Following the lead of Pickwick, mills started sprouting up in this region in the 1850s. The city of Faribault opened the first mill, albeit a sawmill, in 1856. Also during that year, John North and Felix Collett built a mill in Northfield, Minnesota. One year later, the structure was converted into a grist mill by Henry Riedell. The following year, John and George Archibald constructed a stone mill near Dundas, Minnesota. In 1862, two more mills, the Polar Star Mill and the Faribault Mill, were erected in Faribault, Minnesota.[3] The milling industry was in full swing and was rapidly advancing upon Minneapolis and St. Paul.

By the dawn of the 1870s, Minneapolis embarked upon its meteoric rise to flour milling dominance.[4] As the Twin Cities marched onward with its industrial growth, the structural revolution taking place in Chicago seemingly failed to persuade Minnesota builders that heavy timber construction should be avoided in favor of new technology. Some observers voiced strong opinions that brick and timber were perfectly acceptable and were safe so long as common sense fire safety provisions were implemented.[5] Other observers approached Chicago's metal framework revolution cautiously, with one stating:

---

[3] Paul R. Fossum, 'Early Milling in the Cannon River Valley.' *Minnesota Historical Society*, (Northfield: Carleton College, 1930), pp. 274–275.

[4] Alison Watts, 'The Technology That Launched a City.' *Minnesota Historical Society*, (Minneapolis: Minnesota History, 2000), p. 88.

[5] J. A. F., 'Checking the Fire Fiend.' *Scientific American*, XXXIV.17 (1876), p. 261.

> Iron, I mean cast iron, absurd as the statement may appear, will not resist as much heat as good sound oak timber of the same dimensions. Fire expands the iron and warps it, and it breaks very easily. Indeed, if oak timber should be treated by any of the processes of liquid silicate, it may be considered almost a fireproof material compared with cast iron. As for stones suitable for building purposes, as I told you before, there are few that are fireproof, though some approximate the necessary conditions, and, except in severe conflagrations, may be generally depended upon. Granite, marble, and sandstone are not to be trusted, as they soon perish by exposure to the heat, as has been shown a thousand times. But I am strongly in favor of liquid silicate as a preparation for wood to be used for building purposes. My attention was directed to this material some years since, but I have not had an opportunity to investigate the subject fully. I believe, however, that it merits more attention than any other suggestion that has been made public, and may yet prove one of the most practical solutions of the question of non-combustible construction that has yet been offered. Whether this or some other process for making wood non-combustible is the more desirable, I am not prepared to say. I am, however, decidedly of the opinion that any process by which wood can be rendered non-inflammable at a reasonable cost would not only be an inestimable blessing to the public, but its use should be rendered imperative by law....
>
> I looked into that subject at Chicago with much interest. Now, it is very hard to make an absolutely fireproof building; but I believe that a building, properly constructed of bricks that are well made, and of iron or non-combustible timber, protected by fireproof shutters and door, will resist the fiercest conflagration.[6]

These are sentiments that Minneapolis builders seem to have latched onto. Minneapolis itself was undergoing a period of profound population growth. Its bustling milling industry propelled the city's population to grow from 5,809 in 1860 to 129,200 in 1885.[7] By 1874, the Washburn A Mill had been completed and stood as the premier flour mill in the country, and flanked by a rapidly developing heavy timber-based milling district, Minneapolis quickly solidified its position of budding industrial might. From 1870 to 1880, Minneapolis built 17 flour mills, which allowed the city's annual flour output to climb from 850,000 barrels in 1875 to 7 million barrels in 1889.[8]

Yet an event in 1878 threatened to undo all of the progress that had been made, and possibly cause the ultimate extinction of heavy timber construction. Dust had long been known as a significant hazard capable of producing incredibly powerful explosions and devastating fires.[9] On 2 May 1878, the vaunted Washburn A Mill and a significant portion of the Minneapolis milling district, including the heavy timber Thompson & Hoyt and Humboldt mills, were reduced to rubble or heavily damaged following a dust explosion. An investigation revealed that the ignition of combustible mill dust generated and allowed for the expansion of flammable gases, resulting in the death of 18 people.[10] One eyewitness account vividly recalled the disaster, stating:

> From my own point of observation, which was about a mile distant, but two distinct explosions were heard; others nearer heard three, the first not as violent as the other two; while those nearer still heard in addition a sound which they described as a succession of sharp

[6] H. J. Ramsdell, 'Fireproof Materials.' *Scientific American*, XXV.25 (1871), p. 385.

[7] Watts, p. 94.

[8] Watts, p. 94.

[9] 'Explosions of Flour Dust.' *New York Evangelist*, 23 January 1879, p. 7.

[10] 'Terrible Explosion.' *New England Farmer, and Horticultural Register*, 11 May 1878, p. 2.

hisses, resembling the sound of burning gun-powder. Those observers to the windward, whose attention was arrested by the light produced, beyond the distance of half a mile, heard only one or two reports or failed to hear any report at all. From all the testimony in reference to sound it appears that the blow upon the air was not sufficiently sudden to produce a penetrating sound, but rather a dull, heavy blow, which was not communicated laterally to any great distance.

Burning wheat or flour was smelled for several minutes before the explosion by persons in such a position that the wind would carry the odor to them. Smoke was also seen issuing from what was known as the exhaust flour-dust spout of the Washburn A Mill for several minutes preceding the explosion.

At the instant the explosion occurred all observers agreed that the Washburn A Mill was brilliantly illuminated from basement to attic....[11]

The eradication of several heavy timber structures and the possible annihilation of Minneapolis' booming milling industry seemed to spell the demise of heavy timber. However, despite the calamity caused by what was one of the most catastrophic events ever experienced by a heavy timber structure, the region did not lose faith in the construction type. Instead of embracing the revolution emanating from Chicago, the Washburn A Mill was rebuilt in heavy timber (Fig. 4.2). By 1879, this new marvel of heavy timber construction was rising. An article from the *Scientific American* recounted the building's erection, stating:

Among the giant mills which rise on every hand about the milling districts of Minneapolis, the great "Washburn A," says the *Pioneer Press*, looms up conspicuously. Beside it the Humboldt and the Pettit and the Arctic and others in that vicinity look like pygmies. From the canal way back to Second street, a distance of 250 feet, and with a frontage of 100 feet on the canal, the solid walls of limestone are slowly rising under the skillful guidance of Mr. McMullen, the builder of the "B" Mill. These are to be carried to the height of eight stories, thus making the building not only the largest mill on the ground, but the highest of any in the city, for the distance from the level of the canal to the capstone will be 114 feet. To gain an idea of its size one needs to walk about it, both outside and in. The railroad which runs through the building on the second story seems to take up but little room, and yet think of a train of cars passing through any other of our public buildings, how much room would there be left besides? The height of the basement story seems considerable to look at, yet one gets but an imperfect idea of the vast amount of space until he is told that the western half, which is to be used for storing, will hold 100,000 bushels. He can get another idea of its size by figuring the area, when he will be astonished to discover that there are inclosed [sic] 8,850,000 cubic feet.

How much flour this monster is to turn out when completed is a secret which Mr. Washburn keeps to himself.[12]

Upon its completion, the new Washburn A Mill was the pearl of Minneapolis, and it had been outfitted with three particular technological innovations that not only streamlined production, but increased safety, notably the middlings purifier, the gradual reduction process, and the Behrns millstone exhaust system. In short order, Minneapolis sat as the undisputed ruler of flour production.[13] However, disas-

[11] S. F. Peckham, 'On the Explosion of the Flouring Mills at Minneapolis.' *American Journal of Science and Arts*, 16.94 (1878), p. 301.

[12] 'The Largest Flour Mill in the World.' *Scientific American*, XLI.19 (1879), p. 291.

[13] Watts, p. 87.

Fig. 4.2 The rebuilt Washburn A Mill in the 1960s (Photograph courtesy of the Library of Congress)

ter struck again in 1881, which saw a massive fire destroy numerous heavy timber mills, notably the Minneapolis Flour Mill, Clapp Woolen Mill, Empire Flour Mill, Alaska Mill, and Excelsior Flour Mill, among others. Again, the building response featured resounding and continued support for heavy timber, with most of these structures being rebuilt.[14]

In the years that followed the blaze, additional massive heavy timber building sprouted up in rapid succession throughout the city (Fig. 4.3). In 1880, the beautiful Crown Roller Mill, with its finished facades and tin mansard roof, was completed.

[14] Anfinson, p. 327.

Fig. 4.3 The Minneapolis Milling District in the first half of the twentieth century (Photograph courtesy of the Library of Congress)

Between 1880 and 1906, the Dann Barrel Company built several structures to both produce barrels and bags for flour. In 1879, the Standard Mill was built and later enlarged in 1881. The vaunted Washburn A Mill Complex itself underwent renovations and additions in 1884–1885, 1894, 1906–1908, 1911, and several more throughout the twentieth century.[15]

Outside of the Upper Midwest, heavy timber construction found one last bastion in St. Louis, Missouri. St. Louis' industrial development generally appears to have taken shape in the latter stages of heavy timber construction's lifespan (Fig. 4.4). The Cupples Warehouse District in downtown St. Louis is a prime example of this rather late boom of local industrial might. The Cupples Warehouse District is a three-block corridor composed of ten warehouses all ranging from five to seven stories in height and made of red brick, all designed for warehousing and distribution purposes. Only the 1917 building deviated from heavy timber construction, instead being of reinforced concrete construction.[16]

[15] Penny A. Petersen, and Marjorie Pearson, *Architecture and Historic Preservation on the Minneapolis Riverfront*, (Minneapolis: Hess, Rose, 2007), pp. 39–43.

[16] Cynthia Longswisch, *National Register of Historic Places Inventory-Nomination Form: Cupples Warehouse District*, (St. Louis: Landmarks Association of St. Louis, 1985), p. 2.

Fig. 4.4 The 1300 Washington Avenue building, an 1899 Eames & Young design and exceptional example of the ornate style seen in St. Louis, Missouri (Photograph courtesy of public hall, in accordance with GNU Free Documentation License)

The Cupples buildings were modeled after an 1889 Eames & Young design, which garnered acclaim both for its architectural expression, dominated by a lightly ornamented facade and articulated with a tri-part horizontal division with continuous arcaded piers and recessed spandrels, and its sound functional and mechanical identity. Within a year of commencing construction on the complex in 1894, ten heavy timber buildings were in the course of being built.[17] Over the course of their lifetime, the Cupples buildings became regional landmarks that not only bore the weight of the city's industry, but were heralded as having "brought warehouse design to a point where little improvement is possible."[18]

Constructed even later than their fellow St. Louis heavy timber brethren in the Cupples Warehouse District, the seven buildings making up the Crunden-Martin Manufacturing Company just outside of downtown aided in propelling St. Louis to its position in the late-nineteenth century as the nation's leading distribution point for wooden and willow wares. Built between 1904 and 1920, the Crunden-Martin buildings are all square and trapezoidal red brick structures ranging in height between five and seven stories. What truly makes these structures unique is that the first five of these structures, completed from 1904 to 1912, were erected decades after the introduction of more advanced and cost-effective technologies. As such, they serve as the last known dedicated heavy timber construction effort in the

[17] Longswisch, p. 9.

[18] Longswisch, p. 9.

Midwest. Moreover, the heavy timber contingent has endured the same strains and survived right alongside their two reinforced concrete siblings.[19]

By the middle of the twentieth century, as air and road transportation rose to supplant traditional freight transportation services such as river barges and railroads, the mighty Midwestern heavy timber buildings that principally sat on waterways and rail junctions faded into disuse. Likewise, the general identity of the American economy began to shift away from traditional manufacturing and into services. These developments, when taken together with resource requirements to ensure that such aged structures remained tenable, did well to usher in the ultimate demise of heavy timber construction in the region. In fact, by the late 1960s, only five of Minneapolis' once 20-strong contingent of heavy timber mill buildings remained.[20]

[19] Matthew Bivens, *National Register of Historic Places Registration Form*: *Crunden-Martin Manufacturing Company*, (St. Louis: Landmarks Association of St. Louis, 2004), pp. 5–12.

[20] Robert M. Frame, III, 'Mills, Machines, and Millers.' *Minnesota Sources for Flour-Milling Research*, (Minneapolis: Minnesota History, 1978), p. 152.

Chapter 5
The End of Heavy Timber Construction

Heavy timber construction had dominated the American landscape for the better part of a century before it eventually lapsed into obsolescence. While heavy timber buildings continued to be built around the country throughout the late-nineteenth century, notably in its manufacturing centers, the dawn of the twentieth century saw the beginning of the time-tested construction type's phasing into obscurity. A number of factors played critical roles in heavy timber's eventual downfall.

While the previously mentioned developments in alternative fire-resistant building developments in Chicago did not immediately persuade other cities and regions to abandon heavy timber construction, later structural design developments emanating from Chicago can be directly attributed as preparatory groundwork for the demise of heavy timber construction. In 1874, Chicago suffered yet another devastating fire; this time the blaze swallowed 12 blocks in downtown,[1] laying waste to the Type III buildings that populated the area. With insurance companies still reeling from the gargantuan losses of the Great Fire, they applied tremendous pressure on both regulators and designers to require the establishment of even more steadfastly fire-resistant buildings.

In 1872, the state of Illinois had passed the Cities and Villages Act, which effectively granted local government authorities increased powers in the regulation of the health, welfare, and general safety of a given municipality's residents.[2] Three years later, a renewed push was made to strengthen and codify the city's building and fire safety standards.[3] On 11 March 1875, an article in the *Chicago Tribune* commented on the proposed code reform, stating:

> We hope, therefore, it will be passed with this clause (1), that the limitation may be passed upon by the Supreme Court once and for all, and (2), that, if it be not unconstitutional, Chicago may thus enjoy a protection without which all the fire engines in Christendom will

[1] Rayfield, 1997.

[2] Joseph P. Schwieterman, Dana M. Caspall, Jane Heron, *The Politics of Place: A History of Zoning in Chicago*, (Chicago: Lake Claremont, 2006).

[3] Schwieterman et al., 2006.

J. Heitz, *Fire Resistance in American Heavy Timber Construction*,
DOI 10.1007/978-3-319-32128-8_5

> not save us from disastrous conflagrations....The existing fire ordinance does not supply the need in any sense. In the first place, it is altogether too general and too vague in its provisions....[4]

Chicago continued to make great use of iron, until its more durable and popular cousin, steel, made its way to the scene. It was in Chicago that the use of metal framing in American construction came into its own. Chicago was still a city in the process of accommodating tremendous growth. In fact, in 1880, a quarter of an acre of land in downtown fetched a price of approximately $130,000, and by 1890, that figure had ballooned to an astronomical $900,000.[5] For this reason, Chicago-based builders and business owners found that it was far more cost effective and outright efficient to build upward, rather than outward, and the king's ransom it would take to build a suitable heavy timber industrial building or complex was simply not feasible.

In the years following the Great Fire, Chicago, with space and financial limitations providing the impetus, ushered in what became the "commercial style," a building design system that adapted bridge building techniques for structures in an attempt to create vertical building space. These structures shifted away from heavy load-bearing masonry walls and toward a load-bearing structural frame. Doing so allowed builders to bypass the traditional requirement of building an even more massive base to support such magnificent vertical loads. The commercial style allowed for rapid construction, lower labor costs, and increased floor and window space, while still relying on traditional materials such as masonry and iron. All of this translated into increased profitability. The development of this style allowed builders to circumvent the typical height limitations of heavy timber, which often maxed out at around five stories in height.[6] The Leiter Building (Fig. 5.1), which was completed in 1879 and occupied 200–208 West Monroe Street in downtown Chicago, was one of the first structures to successfully demonstrate the commercial style.[7]

Continuing to evolve at a breakneck pace, Chicago ushered in skeleton frame construction, which proved to be a massively popular evolution in fire-resistant building construction. The genesis of the skeleton frame allowed Chicago to not only capitalize on astronomical land prices in the business district, but also allowed for Chicago to quickly erect a number of buildings that reached heights yet unseen in the United States' developed centers. As the birthplace of skeleton frame construction, Chicago quickly became the leader in the design of permanent and robust fire-resistant structures, and its skyline was quickly filled with skeleton frame structures which are best described as buildings in which all interior and exterior loads are carried to the structure's foundation by a metal or reinforced concrete framework, either separately or in combination.[8]

---

[4] 'The Proposed Building Law.' *Chicago Tribune*, 11 March 1875.

[5] Rayfield, 1997.

[6] Rayfield, 1997.

[7] Randall, p. 16.

[8] Randall, pp. 14–16.

Fig. 5.1 The Leiter Building in Chicago (Photograph courtesy of Oklahoma University)

The first dedicated effort to build a true skeleton frame building came in 1883, when William Le Baron Jenney penned the plans for a ten-story office building to serve as the headquarters of its namesake, the Home Insurance Company (Fig. 5.2). In December of 1884, Jenney's Home Insurance Company Building was under construction, yet despite this magnificent structure not yet having been completed, public calls for continued erection of the skeleton frame type of building grew even louder.[9]

Prominent architect, Frederick Baumann, published a leaflet arguing forcefully for the adoption of skeleton frame construction, stating, "Occupants seek convenience, secureness, and light, all of this, of course, combined with a shine of elegance." Baumann further argued that skeleton frame construction could provide all of that due to the type's speed and economy of construction, its stiffness, and it being "light—the most indispensable desideratum with a building is procured even in the lowest, most valuable stories, where otherwise the necessarily broad piers would be a hinderment."[10]

In 1886, the city of Chicago passed an ordinance that amended its building code. The new ordinance required that any new buildings in excess of 90 ft in height had

[9] Randall, pp. 14–16.

[10] Randall, p. 16.

Fig. 5.2 Jenney's Home Insurance Company Building in the first quarter of the twentieth century, which also gives a great view of downtown Chicago's architectural identity (Photograph courtesy of *Construction Week Online*)

to be composed of noncombustible materials.[11] The lone exceptions to this requirement were the floors and trim; however, even these facets had to rest upon concrete or some other sort of noncombustible material. By the dawn of the 1890s, when Chicago had seen some $257 million infused into its built environment, $59 million of which came in 1890 alone and resulted in 50.5 new miles of building frontage,[12] the era of concrete and steel had arrived in the "Windy City."

All across the city, colossal buildings were sprouting up. The Depression of 1893 certainly slowed building projects in general, but Chicago quickly overcame the effects of economic turmoil.[13] Buildings utilizing steel skeletons became common sights in Chicago. An article from an 1895 edition of *Engineering News* commented on Chicago's erection of tall buildings, stating:

> The construction of enormously high office buildings with frameworks of iron and steel carrying the exterior and interior walls and partitions, has become an established feature in nearly all large American cities. This style of construction originated in Chicago, in its practical application, at least, and that city has at the present time more buildings of steel skeleton type than have all other American cities together.[14]

As the twentieth century began, fears were raised about the long-term viability and survival of metal framework. In 1902, a City of Chicago committee investigated

---

[11] Randall, p. 13.

[12] Randall, p. 14.

[13] Randall, p. 18.

[14] Randall, p. 14.

several metal-framed buildings constructed in the 1880s and 1890s, notably the Home Insurance Company Building, Rand McNally Building, Caxton Building, Great Northern Hotel, and the Masonic Temple.[15] After extensive analysis, the committee commented:

> All the first buildings constructed with steel frame will endure indefinitely, if reasonable care is exercised to protect the structural material from water. It is true that in those days little care was taken to protect the metal from corrosion as compared with the care ordinarily taken in these days, but in spite of that fact the metal shows clearly that the ordinary hand terra-cotta fire proofing used in covering the columns has protected them from rust wonderfully well, and that the concrete around the grillage beams has likewise served its purpose in a perfectly satisfactory manner.[16]

By the late 1890s, steel and reinforced concrete framing had found a strong foothold in Chicago, and what would become the models for Type I and Type II construction became a favored standard for building fire-resistant structures. Even world-famous architect, Frank Lloyd Wright, designed such structures, notably his 1902 E-Z Polish Factory (Fig. 5.3) and his 1904 Unity Temple.[17] A new era was rising, one that saw the use of heavy timber construction rapidly disappear.

While it is true that heavy timber buildings continued to be erected across the country in the early portion of the twentieth century, a number of reasons ultimately

Fig. 5.3 Frank Lloyd Wright's 1902 E-Z Polish Factory (Photograph courtesy of James N. McNally)

---

[15] Randall, p. 18.

[16] Randall, p. 19.

[17] Randall, p. 20.

led to its demise as the uncontested dominant fire-resistant construction type. Chief among the contributing factors for such a downfall was the input costs themselves. The nation's massive population and infrastructure expansion in the early-twentieth century placed incredible demand on timber supplies, which, by the 1920s, witnessed on average some 250 million trees accounting for 23 billion cubic feet of wood, harvested each year. Railroads alone were utilizing 130 million timbers for ties each year, while the development and maintenance of the nation's telephone and telegraph systems required the annual felling of five million trees to make enough of the required poles.[18]

Meanwhile, the prolific steel and concrete industries had been able to produce an abundance of cost- and labor-effective construction materials. The use of reinforced concrete allowed for new architectural possibilities that exceeded the capabilities of heavy timbers, which by then were in exceptionally short supply. Single spans in a heavy timber building were commonly 14–16 ft; the possibility of 18-ft spans existed, but it took incredible amounts of time and money to procure such large and rare timbers. Additionally, it was critical for structural timbers to be properly seasoned before installation, creating yet another time delay in the overall construction process. With steel and reinforced concrete, spans ranging from 22 to 24 ft became commonplace. Furthermore, these two materials when combined could create a girder-less slab, which drastically cut down on the materials necessary to form a structural frame.[19]

From what Chicago had experienced, the pronounced need to build vertically in order to achieve economic growth became the overpowering trend in most American cities of the period. It was a marvel that steel facilitated such a change, as one commentator stated in the late 1920s:

> If you look at the great office buildings of your native or your nearest city, it is mostly limestone, brick, marble and glass that meets your eye. The amazing material that has made these wonders possible is modestly hidden away. Yet if that material were suddenly removed from the vital place it holds in our civilization today, most of our buildings would collapse, our airplanes would fall from the sky, the useless shells of automobiles would line the highways, railroads would disappear and civilization as we know it would come to a sudden, terrible and tragic halt.
>
> That material is steel.[20]

It was the boom of industry, the widespread growth of cities, and the ready access to affordable and less labor-intensive building materials that all coalesced to drive heavy timber construction into effective extinction. Many localities also produced building codes favorable to these new methods of construction,[21] ensuring that the

[18] F. S. McGinnis, 'Use and Conservation of Our Timber Supply.' *Overland Monthly and Out West Magazine*, November 1928, p. 379.

[19] 'Department of Architectural Engineering: Posts and Lintels.' *The American Architect and the Architectural Review*, 120 (1921), p. 205.

[20] Lee H. Miller, 'Towering Steel.' *Youth's Companion*, May 1929.

[21] R. G. Kimbell, 'Architectural Engineering: Building Code Limitations on Floor Areas in Buildings.' *The American Architect and the Architectural Review* 125 (1924), p. 301.

progress of industry was in no way limited. While new innovations in heavy timber construction continued to emerge, such as the invention of reinforced concrete caps for timber posts,[22] the coming extinction of heavy timber construction was unavoidable.

In a land governed by the laws of economics, heavy timber construction, despite its proven performance marked by individuality, lost the battle against quantity production.[23] During this era, with the decline of heavy timber construction, it can be argued that the very nature of American architecture had changed. As one author perfectly expressed:

> On the architect has fallen the problem of so constructing buildings and houses (and even cities) that all movement is smooth and quick and economical of time and energy. Order is the order of the day. For our massed activities cannot be carried on in disorder. This necessity is the root of modern architecture, as functional order is its very essence. And modern architecture has come to look differently from all previous forms, because it is as different as the life it is designed for.
>
> The forces that changed society had their origin in the industrial revolution. Transforming the nature of production and distribution transformed society. So modern architecture has been made possible by the development of modern industry. New materials, methods and technique not only enlarged the possibilities of building, but fundamentally changed its character. Steel and concrete, increasing strength, opened new structural vistas. They gave us freedom of ground plan, and independence from supporting wall. The whole building bulk was loosened and lightened. Space, formerly so inclosed [sic], could now be treated as something free. Walls, no longer bearing the weight of the building, became little more than a sheathing. The windows, gradually enlarged, often came to replace the walls themselves. The whole transition was from the fixed and inflexible to the free and flexible.
>
> The new technique permits us to do things that could not be done before. We can now span larger spaces and greater heights. We build according to our technique, even as the builders of the pyramids did, or the Greeks, or the men of the renaissance achieving the great Gothic. And the architecture of steel, cement, and glass is naturally enough far different from that of stone or wood.[24]

New York City, while implementing skeleton frame steel construction later than Chicago, saw the construction of the Asch Building located at 23–29 Washington Place in Manhattan (Fig. 5.4). Designed by architect John Woolley, plans for the 135-ft tall tower were filed in April of 1900, and the building was completed in January of 1901. It was a modern loft building with an iron and steel skeleton frame, trimmed with brick in a Neo-Renaissance style. The metal structural members were encased in terra-cotta fireproofing, similar to its Midwestern predecessors in Chicago. While lacking a sprinkler system which had been a fairly common feature in large buildings by this time, the building did feature a standpipe system fed by a rooftop water tank that ran the height of the building, to aid in firefighting efforts should a blaze break out.[25]

---

[22] 'Reinforced Concrete Caps for Wooden Posts.' *The American Architect and the Architectural Review*, 121 (1922), p. 236.

[23] Samuel Warren, 'Texture in Portland Cement Stucco: Part III.' *The American Architect and the Architectural Review*, 126 (1924), p. 319.

[24] William Lescaze, 'The Meaning of Modern Architecture.' *The North American Review*, 244 (1937), p. 110.

[25] Gale Harris, *Brown Building (Originally Asch Building)*, (New York: Landmarks Preservation Commission, 2003).

Fig. 5.4 Firefighting efforts being taken against the 1911 Triangle Shirtwaist Factory fire in the Asch Building (Photograph courtesy of the United States Department of Labor)

The Asch Building was a technological marvel, particularly in its construction time. Such a formidable tower had been erected in less than a year, a feat that no heavy timber building of any size could compete with. However, despite the structure's identity as fire resistant, it was the home of one of the deadliest fires in American history. On 25 March 1911, a fire erupted on the eighth floor in the Triangle Shirtwaist Factory. The FDNY had brought the blaze under control within 25 min after the sounding of the first alarm; however, 146 workers died in the fire.[26] Essentially, this was the first large fire in this new type of construction, which proved that the building type was not necessarily superior to heavy timber in its ability to slow down a blaze, and that fire spread is largely due to the specifics of the

[26] Harris, 2003.

Fig. 5.5 The Baker and Hamilton Building in San Francisco (Photograph courtesy of Christopher Ver Planck)

fire load within the building. This tragic event showed that while the building could survive an intense blaze, with little structural damage, the protection of occupants was a different matter.

Stretching all the way across the country to San Francisco, California, heavy timber construction's lifespan took a bit of a different trajectory. In a most unusual fashion, San Francisco, one of the only principal industrial centers on the West Coast at the time, built heavy timber and skeleton frame buildings side by side. During the period of 1894–1929, but principally between 1906 and 1918, this locality opted to employ both construction types. San Francisco alone houses several of these curious structural pockets, such as Northeast Waterfront District (Fig. 5.5), the South End Historic District, and the Dogpatch Historic District. The largest and aptly named Showplace Square Heavy Timber and Steel-Frame Brick Warehouse and Factory District in San Francisco, which houses at least 16 of these structures,[27] stands as a testament to simultaneous construction of these distinct types, perhaps proving that while heavy timber construction may have been largely replaced by more modern systems, its merits in the realm of fire resistance were still well known.

What is most interesting is that even well into the era of steel and concrete, builders in California steadfastly held onto the benefits of heavy timber construction and the inherently stout characteristics of large-dimension timbers. While later shifting away from heavy timber construction, the type's laboriously and materially intensive nature does not appear to be the primary motivator for the region's change in building identity. What ultimately seems to have forced the abandonment of heavy

[27] Christopher Ver Planck, District Record—New Wholesale District, (San Francisco: Department of Parks and Recreation, 2009), pp. 3–4.

timber construction and steel-frame masonry construction in this era was the ever-present danger posed by earthquakes. At the time of their construction, these buildings both performed ably under the stresses imposed by earthquakes; however, by the time of California's industrial boom in the second decade of the twentieth century, reinforced concrete and corrugated steel had become readily available and firmly replaced the more elderly building types.[28]

Minnesota's Twin Cities area, Minneapolis and St. Paul, the Midwestern juggernauts of heavy timber construction, also continued with the use of heavy timber well into the era of concrete and steel, despite the decades-old structural revolution that had taken place in Chicago. It was not until well into the first decade of the twentieth century that Minnesota started building in steel, with widespread structural use of concrete coming in significantly later.

As of 1906, Minneapolis, which had long been one of the great bastions of heavy timber construction, had continued to follow its rich tradition, seeing the completion of the Butler Brothers Warehouse, an eight-story heavy timber building located at 100 6th Street North. Yet, two years later in 1908, the massive Ceresota Building was completed in the heart of Minneapolis' heavy timber district. The Ceresota Building (Fig. 5.6), constructed out of brick and steel, features a base of 7 stories

Fig. 5.6 The Ceresota Building in Minneapolis (Photograph courtesy of *The Minneapolis Journal*)

[28] Ver Planck, p. 7.

Fig. 5.7 The steel-framed Chaska Sugar Factory in the early-twentieth century (Photograph courtesy of the Carver County Historical Society)

with a tower rising to 11 stories.[29] This lone building rose to heights difficult for heavy timber to achieve. Even in suburban and rural Minnesota, in a small town such as Chaska, the end of heavy timber generally coincided with that of Minneapolis. While this small and typical suburb had rebuilt and built the heavy timber Guardian Angels Church and Mill Building in 1902 and 1903, by 1905, the first steel beams and bricks were laid for the city's sugar factory (Fig. 5.7).[30]

By the close of the 1920s, new construction in heavy timber had ceased, and while its heyday had arguably ended a few decades prior, many of these giants continued to be utilized well into the latter half of the twentieth century. While many have long since been abandoned, some have found new life. In both respects, these magnificent monoliths serve as bold reminders both of the United States' past and its future, and they stand as unmistakable embodiments of American ingenuity.

[29] Carol Ahlgren, *Staff Report—Ceresota Building* (Minneapolis: Minneapolis Heritage Preservation Commission, 2008), p. 2.

[30] Mollee Francisco, 'How Sweet It Is.' *Chaska Herald*, 20 April 2015.

Chapter 6
The Performance of Heavy Timber Construction

Heavy timber construction was expressly designed to withstand the damaging effects of fire, and for its time, it was a revolutionary concept in fire-resistant building construction. Even decades after its fall from popularity, heavy timber is still regarded as one of the premier fire-resistant construction types. Over the course of its 180-year official history, heavy timber construction has compiled a detailed performance history[1] which unequivocally proves that when adequately maintained and built to code specifications, the building type routinely delivers a high level of real-world fire safety.[2]

Great care was exercised in the development of safeguards against fire, the quintessential eternal enemy of buildings. Over time and through numerous rounds of trial and error, the recipe for heavy timber's success was conceived. This is certainly not a fact to be overlooked, as the building type itself was designed to house the varying and often heavy and volatile fire loads of the whole of the industrial sector. Add to this that fire protection systems that are currently widespread were rare during the early history of heavy timber construction, as such the building had to be robust enough to not only withstand the effects of the fire, but do so for long enough to allow the large number of occupants to escape.

The strengths of heavy timber construction rest upon several hallmark traits. It was determined that when interiors of large spans were framed with timbers of exceptionally large cross-sectional area, extended burn times prior to structural collapse were possible.[3] It was also determined that in order for the aforementioned truth to hold, the timber members themselves absolutely had to possess smooth surfaces and that the building had to be without void spaces or unprotected passageways

[1] Technical Services Division, *Heavy Timber Construction Details*, (Washington, D.C.: National Lumber Manufacturers Association, 1961).

[2] Robert H. White, *Reporting of Fire Incidents in Heavy Timber Structures*, (Madison: Forest Products Laboratory, 1985), p. 1.

[3] James Patterson, *Simplified Design for Building Fire Safety*, (New York: Wiley, 1993), pp. 110–111.

J. Heitz, *Fire Resistance in American Heavy Timber Construction*,
DOI 10.1007/978-3-319-32128-8_6

into other parts of the structure.[4] The ends of girders were often "fire cut" so that if a fire resulted in the collapse of a floor, it would not drag the heavy walls down with it. Protection of the structural members from moisture was also a chief design feature, with many of these buildings incorporating scuppers in the walls to drain off water.[5] Additionally, the existence of fire walls or large fire doors within the structure was also critical so that fire spread would be limited to the area of origin.[6]

Perhaps the true testament to the incredible performance of these structures is that they had survived and were utilized well into the 1950s and 1960s, sometimes without many of the modern fire safety systems that modern codes would require. Additionally, these structures still continued to shoulder the strains of physically bearing the demands of manufacturing, right up until the arguable end of the United States' identity as primarily goods- or manufacturing-based economy. However, the long continuous lifespan of these structures themselves can deteriorate their fire-resistance over time. Some structures have fallen victim to numerous small fires in the same general area, which have weakened the structural members.[7]

What has also been found in heavy timber buildings is that they are quite resistant to structural collapse. The American fire service generally regards them as the second most resistant building type to structural collapse. Under duress, heavy timber construction buildings have proven themselves, due to their extremely massive load-bearing exterior walls, to be unusually stable.[8] It has been found that when fire has erupted within a heavy timber building, the structure is not prone to collapse, albeit catastrophic collapse, until the generally large floor and wall assembly connectors and fasteners, often made of steel or iron, are compromised. To be sure, while structural collapse in heavy timber buildings often takes considerable time to occur, when it does, it often involves large areas and both walls and floors collapsing at once.[9] Additionally, the collapse of upper floors can cause lower floors to collapse in succession, almost as if they were dominos.

It seems almost unfathomable that a building construction type which relies on the use of a dried timber, what was once a living tree in a forest, can be termed "slow burning," yet this physical contradiction deserves its well-earned reputation.[10] The creators of this construction type seemed to have understood this apparent paradox when designing its standards. Aside from the previously mentioned necessary structural features heavy timber buildings were required to possess in order to be truly fire-resistant, a number of practices were instituted that sought to further protect the structures during the inevitable outbreak of fire. Structural designers of the era knew well that for buildings stocked with highly combustible materials and arranged in

---

[4] Patterson, p. 111.

[5] Francis L. Brannigan and Glenn P. Corbett, *Brannigan's Building Construction for the Fire Service*, 4th ed., (Sudbury, MA: National Fire Protection Association, 2007), p. 157.

[6] Brannigan, p. 157.

[7] John Norman, *Fire Officer's Handbook of Tactics*, (Tulsa, OK: PennWell, 2005), p. 389.

[8] Norman, p. 389.

[9] Norman, p. 389.

[10] Brannigan, p. 157.

large open spaces that could reach 30,000 ft^2 and that if fire were to erupt, not only would it be an utmost priority to evacuate the occupants, but any fire not extinguished quickly could rapidly exceed the suppression capabilities of local fire departments.[11]

The first major step in defending these industrial castles from fire was the organization of contingents of a heavy timber building's occupants, which often numbered in the hundreds during the construction type's zenith during the American Industrial Revolution, into a veritable volunteer emergency services organization. Employees would assume various posts within this preemptive emergency organization, and evacuation drills were routinely practiced, although fire drills did not become governed by code until 1911 following the fire at the Triangle Shirtwaist Factory.[12] Some employees served as fire wardens, who were trained to coordinate the orderly evacuation of workers and occupants on each floor of the building. Other employees were organized into fire brigades, and their job was to attack and suppress or otherwise extinguish incipient fires, either buying time for the occupants to escape or putting out the fire altogether. Additionally, there were searchers who would search dressing rooms and restrooms to ensure that any occupants in those areas were safely evacuated. Some employees served as fire alarm box runners, whose job was to transmit a fire call to the local fire department. The final component of this system were the door holders, who would assist fleeing occupants navigate each floor's exits and then close the doors following the full evacuation.[13]

To be clear, fires in heavy timber structures were common occurrences. Sometimes, they were disastrous, such as the 1878 Minneapolis Washburn A Mill, a flour mill, which suffered a dust explosion that decimated the entire structure and generated enough energy to blow the roof assembly several hundred yards into the air.[14] However, events such as these have historically proven to be the proverbial exception to the rule. Far more common was that most of these fires were extinguished by on-site personnel while in the incipient stage. An article from a 1907 edition of the *Quarterly of the National Fire Protection Association* perfectly highlighted not only the more common scene concerning fires in heavy timber buildings, but how well these structures performed under fire conditions when properly maintained and designed, even when significant human error was involved. The piece summarized a fire in a Philadelphia cotton mill and stated:

> Fire occurred on second floor in first break of cords, open mill construction. Fire spread rapidly and opened up about forty-five sprinklers which controlled the fire satisfactorily. Sprinklers were shut off and shortly after engineer and machinist started replacing sprinklers on the second floor. While doing this fire started up again in a bundle of yarn and spread quickly over piles of yarn and bales of cotton. Sprinkler valve was opened up as quickly as possible and probably only a few minutes elapsed but this was sufficient to allow

---

[11] Dunn, p. 148.

[12] Dunn, p. 148.

[13] Dunn, p. 148.

[14] Manju Mittal, 'Explosion Hazard and Safety in Industries Handling Grain Products.' *Journal of Engineering Research and Studies*, IV.3 (2013), p. 2.

> fire to gain headway. Fire spread through small belt hole into third story and altogether some seventy heads opened on the second floor and forty-eight on the third. Fire also burned through roof at one end. It was finally controlled by sprinklers and hose streams.[15]

While many of the performance figures for heavy timber construction had undoubtedly been generated through careful observations of architects, engineers, craftsmen, and investigators, following fires, modern scientific approaches have greatly enhanced our understanding of just how and why heavy timber construction has been such an effective form of fire-resistant construction. Building codes largely only rate the masonry elements of heavy timber buildings. Exterior walls, fire walls, and vertical openings are all required to achieve 2-h fire ratings. Interior bearing walls are required to have 1-h fire ratings.[16]

Perhaps what is most interesting about the fire performance ratings of heavy timber structures is that the wood elements themselves do not carry specific ratings. The Underwriters Laboratories, which is the preeminent fire performance testing organization, opted not to test structural timbers as they have been deemed inherently fire resistant.[17] However, other organizations have undertaken extensive testing to determine just how large timbers perform under fire conditions. What has been found is that a visible insulating char layer develops at around 550° Fahrenheit. It has also been discovered that wood below the char layer can withstand days of continuous subjection to temperatures of 150 °F and hours of exposure to temperatures below 212 °F, without losing their mechanical properties. However, when this layer is exposed to temperatures above 390 °F, the timber experiences significant degradation in its mechanical properties.[18]

With the advent of glued-laminated timbers, or informally known as glulam, coming to dominate both new builds and renovations of heavy timber buildings, even more testing has been on timbers under fire conditions. What these tests have sought to highlight is that no form of construction is "fireproof," that the best one can hope for is to be "firesafe," and that in a worst-case scenario the structure will resist fire impingement long enough to facilitate the safe evacuation of occupants.[19] To that end, it has been found that the modern structure fire averages sustained temperatures ranging from 1290 to 1650 °F and that even steel starts to weaken at 450 °F, losing its structural integrity to point that at 1380 °F; only 10 % of its strength remains. What these tests have concluded is that wood typically won't ignite until it reaches 480 °F and that during a typical fire glulam members, like their sawn cousins, lose about three-fourths of an inch to charring each half-hour.[20]

While heavy timber construction has an almost pristine reputation in successfully dealing with the small fires it was intended to contend with, and while it has

---

[15] Henry A. Fiske, *Quarterly of the National Fire Protection Association*, 1 (1907), pp. 74–75.

[16] *Fire-Rated Systems*, (Tacoma: APA—The Engineered Wood Association, 1999).

[17] Brannigan, p. 159.

[18] Robert H. White, 'Post-Fire Analysis of Solid-Sawn Heavy Timber Beams.' *Structure Magazine*, November 2013.

[19] *Superior Fire Resistance*, (Centennial: American Institute of Timber Construction, 2003).

[20] *Superior Fire Resistance*, 2003.

shown tremendous resistance to incredible fires, the type has also encountered staggering losses during fire events. Certainly, a properly built and correctly maintained heavy timber building presents an environment that allows for the successful combating of flames. However, modifications, abandonment and deterioration, years of flammable solvents soaking into the timber, and the absence of necessary sprinkler and alarm systems, or the possession of fire loads beyond the aforementioned system's capabilities, can and have led to fearsome and tragic fires which have resulted in both fatalities and the reduction of the structure to rubble. To make matters worse, if the blaze is not brought under control in a timely fashion, these massive buildings, often arranged in multiple-building complexes, can breed conflagrations whose powerful updrafts can ignite structures hundreds of yards away (Fig. 6.1).

In the era following heavy timber construction's fall from popularity and its replacement with more advanced fire-resistant technologies, many of these buildings have fallen into disrepair, suffered the effects of aging, or have been repurposed for occupancies and fire loads they were not designed for. Unfortunately, when fire erupts in such buildings, it often ends in catastrophe. While such events were certainly not common outcomes during the apex of heavy timber construction in the United States, they have become increasingly common in the last half-century and serve as a constant reminder that undermining or otherwise disregarding the carefully calculated design features can lead to spectacular fires.

On 14 February 1958, a fire erupted in a six-story heavy timber building with brick walls, measuring approximately 80 ft by 100 ft, and built a hundred years prior in 1858. The building was located at 137–139 Wooster Street in New York City, in a region of Manhattan known as "Hell's Hundred Acres" due to its infamous reputation for killing numerous firefighters in frequent large-scale blazes. The region was dominated by United States Civil War Era structures, many of them heavy timber buildings.[21]

At the time of the fire, this building was filled with rag storage, baled goods, paper rolls, and multiple heavy machines. As such, the weight and fire load within the building were astronomically high. Reports initially indicated an undetermined explosion after the fire had started, and when firefighters began their initial fire attack, the flames quickly spread vertically up five floors via open shafts. As fire service personnel were conducting salvage and ventilation operations, floors two through six suddenly collapsed, killing two members of the FDNY and four members of the fire patrol; however, all seven employees managed to escape unassisted. An investigation later concluded that the great increase in the building's load when the soluble contents absorbed water from hose streams, and the failure of cast iron columns which complimented the street-side bearing wall, led to the fatal collapse.[22]

[21] Christopher Naum, 'Lessons from the Fireground-1958: Wooster Street Fire & Collapse.' *Fireground Leadership*, 16 February 2015.

[22] Naum, 2015.

Fig. 6.1 A photograph of a 1925 fire at the heavy timber structure in Washington, DC, which vividly depicts the resources needed to combat fully developed heavy timber structure fire (Photograph courtesy of the Library of Congress)

On the night of 27 December 1983, the Buffalo, New York, Fire Department responded to what can only be called its worst event.[23] Fire service personnel responded to a reported propane leak in a five-story heavy timber structure located on the junction of North Division and Grosvenor Streets and nestled into a quiet

[23] Gene Warner, '30 Years Later, Survivors Recount Horror of Propane Blast.' *Buffalo News*, 21 December 2013.

residential area. The first on-scene engine company and battalion chief reported "nothing showing." Less than a minute later, a massive explosion that could be heard some 15 miles away leveled the building. The force of the explosion had blown a fire truck across the street into houses and ignited a blaze that also destroyed approximately 30 homes. During the subsequent operations, all commanded by a district chief who had been impaled in the neck by a 5-in. stake, it was revealed that four complete blocks had been destroyed and buildings up to a half mile away had been damaged. When the dust had settled, six people had perished, including five in the Buffalo Fire Department's Ladder 5 Company, and some 70 were injured.[24] An investigation quickly reported that the explosion was the result of unapproved storage of a 500–1000-gal propane tank that was ignited by a wood-burning stove in the basement.[25]

On 12 March 1987, a massive blaze reduced two warehouse complexes to ruins and led to the deaths of three Detroit firefighters.[26] The North Complex was an exclusively heavy timber construction and was built in phases between 1900 and 1920. The complex had been abandoned for a significant period of time and featured a three-story warehouse and attached four-story warehouse. The three-story buildings had numerous vertical openings, and while the two structures were separated by a masonry firewall, many of the fire doors were missing. Additionally, while the buildings had sprinkler systems, they were inoperative. The four-story structure also housed piles of discarded clothing and what may have been oil soaked rags, which accelerated the fire's growth. Likewise, the four-story building was also full of highly combustible materials.[27] As the North Complex blaze grew, it ignited the similarly composed and likewise abandoned mix of heavy timber and steel-frame warehouses in the South Complex which were filled with paper stock. During the initial fire attack, crews were caught in a flashover, which led to the death of one firefighter. Two other firefighters were killed 2 h into the battle, when performing overhaul operations, and a wall and several floors collapsed on top of them. An investigation concluded that the fire was a result of arson and that the unusually high fire load and general state of disrepair of the abandoned structures were contributing factors.[28]

On 3 December 1999, the Worcester, Massachusetts, Fire Department responded to a call of smoke showing from an abandoned six-story heavy timber cold storage warehouse built in 1905, with an addition dating to 1912 (Fig. 6.2). Upon their arrival, the fire had been burning for anywhere between 30 and 90 min. Over the course of initial operations which lasted 73 min, firefighters conducted a primary

[24] Warner, 2013.

[25] Lindsey Gruson, 'Illegally Stored Propane Is Blamed for Blast That Killed 6 in Buffalo.' *The New York Times*, 29 December 1983.

[26] J. Gordon Routley, *Detroit Warehouse Fire Claims Three Firefighters*, (Emmitsburg: United States Fire Administration, 1987), p. 1.

[27] Routley, pp. 2–3.

[28] Routley, pp. 7–10.

Fig. 6.2 The 1999 Worcester Cold Storage Warehouse Fire that killed six firefighters (Photograph courtesy of the Centers for Disease Control and Prevention)

search to locate potential occupants.[29] Unbeknownst to firefighters, the interior walls had been covered with anywhere from 6 to 18 in. of asphalt-impregnated corkboard and 4 in. of highly flammable polystyrene.[30] The result was incredibly volatile fire conditions within the highly altered and dilapidated building, which led to six firefighters getting incapacitated and perishing in the blaze.

The Worcester Cold Storage Warehouse Fire is one of the quintessential heavy timber fires that proves how alterations and disrepair can lead to catastrophic outcomes in these massive structures. It also shines incredibly valuable light on how simple abandonment and the presence of squatters can doom these buildings and all those who step foot in them. Court records, in which a criminal indictment was filed against the squatters, place the bulk of the blame for the deadly fire at the feet of human beings as opposed to the many contributing facets of the building itself. As documents from the criminal proceeding stated:

> Evidence presented to grand jury that criminal defendants had started a fire and then increased the risk of harm from that fire without taking adequate steps to report it to the proper authorities, that the defendants' choice not to report the fire was intentional and reckless, that the defendants' conduct was the cause that necessarily set in operation the factors that caused the deaths of six fire fighters who fought the fire, and that the inability

[29] Richard W. Braddee, *Six Career Fire Fighters Killed in Cold-Storage and Warehouse Building Fire-Massachusetts*, (Washington: National Institute for Occupational Safety and Health, 2000), p. 1.

[30] Braddee, p. 4.

> of the fire fighters to navigate the building or estimate the true caliber of the danger was foreseeable to those in the defendants' position as a part of the normal risks of combating a fire, was sufficient to constitute probable cause that the defendants had committed the crime of involuntary manslaughter and to justify a return of indictments for that crime.[31]

A final case study that highlights the troubles vacant heavy timber buildings face, yet reinforces the notion of their uniquely robust nature, was a fire that broke out on 8 December 2011 in St. Louis, Missouri. The fire gripped the former Crunden-Martin Manufacturing Company Building, a six-story riverfront heavy timber building, but astonishingly, the St. Louis Fire Department was able to keep the blaze from spreading to nearby structures. Perhaps the most interesting part of this event was that fire department operations lasted for 18 h, and yet the building remained standing.[32]

Over a century of real-world fire performance tells us many things. It cements heavy timber construction's place as among the most fire-resistant structures in existence. However, it also clearly shows that deviations from, or deficits of, its required design principles can significantly undermine their performance under fire conditions. When left vacant or idle, they can become the targets of vandals and deterioration.[33] That while resistant to structural collapse, when such an event occurs, it is a catastrophic structural failure.[34] We have learned that while these structures do not readily allow fire to take hold, when they do, the conflagration is so great that nearby exposures are usually destroyed as well.[35] What has also become crystal clear is that when fire in a heavy timber building is allowed to become fully developed, it will take tremendous resources in terms of personnel[36] and water, with an excess of 3000 gal per minute of flow, to subdue a single fully involved floor.[37]

However, the lessons we have learned in the past half-century with regard to the performance of strengths and limitations of heavy timber buildings were known largely before the construction type's decline. A 1921 *Outlook* article tactfully sums up many of the salient points we have long known that could avoid the visitation of disaster upon heavy timber buildings. The article stated:

> The annual per capita loss from fire in the United States is at least five times the per capita loss of any other country. More fireproof construction, a larger use of sprinkler systems and chemical extinguishers, a more general understanding of the danger arising from volatile oils, a more searching inquiry into the causes of fires, and the surer and more drastic punish-

---

[31] Commonwealth vs. Thomas S. Levesque, Superior Court of Massachusetts, March 2002.

[32] Christine Byers, 'St. Louis Blaze That Burned 18 Hours Now Out, Under Investigation.' *St. Louis Post-Dispatch*, 9 December 2011.

[33] Willis Group Holdings—North America, Strategic Outcomes Practice, *Managing Risks in Vacant and Idle Buildings*, (Chicago: Willis, 2010).

[34] Christopher J. Naum, *National Fire Fighter Near-Miss Reporting System May-Structural Collapse*, (Washington: Command Institute, 2015), p. 11.

[35] Mike Coffey, 'Building Construction: How It Effects Your Firefighting.' *Carolina Fire Journal*, (2011).

[36] Michael L. Smith, 'Heavy Timber Construction: What Firefighters Need to Know.' *Firehouse Magazine*, 1 July 2000.

[37] Thurston County Washington, *Fire Flow Requirements 2013*, (Olympia: Thurston County, 2013).

ment of fire-bugs will do much to cut down this shameful loss. It will not be brought down to a reasonable ratio, however, until our citizens take it upon themselves to overcome their National inclination towards carelessness and reckless risks. When the danger from a carelessly dropped match or a single pile of combustible refuse may result in a loss of millions of dollars, it seems as though the need for caution and care were too obvious even to need mention. It is the obvious danger, however, which is most often overlooked. As long as Americans continue to tax themselves by their present annual fire loss, they are not in a position to cast stones at those who choose to live in the shadow of active volcanoes.[38]

[38] 'Fighting Fire.' *Outlook*, 5 October 1921, p. 163.

Chapter 7
The Resurgence of Heavy Timber Construction

In the second half of the twentieth century, heavy timber construction seemed to have been all but left for dead. It had been deemed too costly, too labor-intensive, and simply too old-fashioned. The American economy had shifted, and heavy timber was literally left behind, often abandoned and left to suffer either a slow and inevitable fate marked by decay, or face destruction by fire or demolition. However, late in the twentieth century and certainly in the twenty-first century, the notion of once again harnessing the possibilities of heavy timber has been rekindled in earnest.

Repurposing is a common sight in many localities. Formerly abandoned heavy timber structures have been seen as beneficial and practical solutions for the common and widespread problem that is urban sprawl and urban decay. City planners across the nation have often found themselves staring at such monoliths and have entertained notions of revitalizing dying urban centers by reviving such structures.[1] In many districts where heavy timber-based industry once powered the neighborhoods, the streets are dark and barren, and economic benefit, much less growth, was nonexistent. To solve this problem, conservation specialists, builders, and government officials have sought, with success, to turn vacant heavy timber buildings into mixed-use properties that will not only attract residents, but will attract both businesses and customers as well.[2]

One of the early examples of such repurposing or revitalization of elderly and vacant heavy timber buildings occurred in one of the Midwestern bastions of the construction type, Minneapolis. In 1986, a dedicated effort was launched to rehabilitate two old heavy timber mills. The goal was to give these structures a second lease on life by making them attractive within a new urban downtown environment. These buildings were to have plazas, underground parking, and restaurants and even feature the restoration of their original water raceways. The old Standard Mill

[1] Lisa Rauschart, 'Cover Story: Old Buildings 'Repurposed' as Unique Housing' *The Washington Times*, 6 September 2012.

[2] Rauschart, 2012.

J. Heitz, *Fire Resistance in American Heavy Timber Construction*,
DOI 10.1007/978-3-319-32128-8_7

Fig. 7.1 Minneapolis' Crown Roller Mill after conversion (Photograph by Susan G. Lesch, used in accordance with Creative Commons Attribution 2.0 Generic License)

Building became the luxurious Whitney Hotel, while the Crown Roller Mill (Fig. 7.1) was transformed into a bustling office complex.[3]

In Lewiston, Maine, the Bates of Maine Woolen Mill, which saved from a fate of destruction by the Environmental Protection Agency of all groups, among others. The Bates of Maine Woolen Mill, which dates to the American Civil War Era, had provided the uniforms for the Grand Army of the Potomac, and, until ultimately losing out to intense foreign competition, was the state's single largest employer through the 1960s. Following the mill's cessation of operations in the early 1990s, the city was faced with a monumental task in dealing with both the environmental and financial implications of demolishing the 1.2 million-square foot building. After several years of struggling, a cooperative of private investors and local government managed to secure some $41 million dollars in funding to repurpose the gigantic structure. The building was then transformed into a thriving multi-occupant office complex complete with fitness and day care centers for the buildings' many employees. As of 2003, a quarter of the building's square footage had been leased out, creating an estimated 1400 new jobs. It was anticipated that when the building was completely overhauled and rented out, some 5000 employees would occupy the structure, quite an economic boom for a city of 29,000.[4]

In Lowell, Massachusetts, arguably the former epicenter of the nation's textile industry, the 500,000-ft^2 Lawrence Mill sat idle, as its once indomitable industrial might had met veritable extinction by the close of the 1970s. Yet in 1987, this giant was repurposed in a rather unique way; it became home to the Middlesex Community

[3] 'Mill District Development Begins.' *Preservation Matters*, 2 (1986), p. 1.

[4] Environmental Protection Agency, *Former Civil War Textile Mill Is Given New Economic Life with the Help$775,000 in Funding from the EPA*, 3 December 2003.

Fig. 7.2 St. Louis' International Hat Company Building (1904) after conversion (Photograph by Bob Patton)

College and the University of Lowell Engineering School.[5] Elsewhere in New England, similar feats were taking place. An old mill along the Blackstone River in Rhode Island was repurposed as a senior living community.[6] In Providence, Rhode Island, the birthplace of heavy timber construction, a former rubber factory became host to a boutique mall and office complex.[7] Even more inspiring was the repurposing of the 20-building Sprague Electric Co. Mill Complex into the Massachusetts Museum of Contemporary Art.[8]

In Missouri (Fig. 7.2), the Cupples Warehouse District, among others, has also undergone a resurgence. While only nine of the original structures remain, St. Louis has faced the problem that not all nine of these buildings are salvageable. In fact, in 2013, demolition began on the Cupples 7 Building. While this lone structure was lost, the remaining eight have been rehabilitated.[9] However, the Cupples 9 Building survives as a triumph of repurposing. After a $30-million-dollar renovation, the 120-year-old heavy timber building has been the home of numerous office occupants.[10] Indeed,

[5] Ken Franckling, 'Historic New England Buildings Increasingly Fill Today's Needs: Reuse of Old Mills, Factories Has Become Economic Phenomenon in New England.' *Los Angeles Times*, 8 December 1987.

[6] Franckling, 1987.

[7] Franckling, 1987.

[8] Franckling, 1987.

[9] Nicholas J. C. Pistor, 'Cupples Complex Shrinks as St. Louis Struggles to Keep Historic Buildings Standing.' *St. Louis Post-Dispatch*, 9 July 2013.

[10] Tim Bryant, 'As Cupples 7 Goes Down, Cupples 9 Thrives.' *St. Louis Post-Dispatch*, 9 July 2013.

Fig. 7.3 Minneapolis' Mill City Museum, in the ruins of the former Washburn A Mill (Photograph by Brady Willette, courtesy of the Minnesota Historical Society)

the thriving of the Cupples Complex is quite a feat given that it very nearly was entirely demolished in 1989.[11]

Perhaps the most unusual and fascinating repurposing of an old heavy timber building is that of Minneapolis' Washburn A Mill, which was virtually destroyed in 1991 by a fire set by vagrants. After halting initial thoughts of simply demolishing the structure, a coalition of local government agencies, preservation organizations, and private industry formed to drastically repurpose the ruins. The result was the Mill City Museum (Fig. 7.3), a nine-story museum which resides within the ruins using both the building's remnants and the picturesque view of the old milling district to inform visitors of the city's rich flour milling history.[12] Great care was taken to complement the ruins and maintain its general ambiance, and while new materials were used for the construction of the museum, the new structure was formatted as to not further degrade the ruins of the Washburn A Mill.[13]

While many heavy timber structures around the nation are being repurposed and renovated to varying degrees, in some rare instances, the old structures are being entirely revived. After several years of having sat vacant, in 2011, Anderson-KM Builders proudly announced that it had commenced work on rehabilitating the

---

[11] Pistor, 2013.

[12] *Preserving America's Heritage: Companion to National Historic Preservation Act*, (Washington: Advisory Council on Historic Preservation, 2006), p. 21.

[13] Brian Schaffer, *Heritage Preservation Commission Staff Report – Washburn A Mill, Mill City Museum*, (Minneapolis: Minneapolis Heritage Preservation Commission, 2008), p. 2.

Fig. 7.4 A photograph of the Faribault Woolen Mill shortly after its renovation and resumption of operations (Photograph by Elizabeth A. Gates, courtesy of Minnesota State Historic Preservation Office)

170,000-ft^2 Faribault Woolen Mill (Fig. 7.4).[14] A 2014 article happily reported that the 1894 heavy timber-hybrid building had returned to producing textiles, while adding both a showroom and executive offices.[15]

The second life of heavy timber construction hasn't simply been limited to repurposing and revivals, relying upon either existing or otherwise foreign technology. In recent decades, a significant effort has been dedicated to pioneering new heavy timber methods and technologies to create a new class of heavy timber buildings. For numerous reasons, the aesthetic value of wood has been a coveted trait by builders, who readily express their fondness for its warmth, distinctive color, and natural grain patterns.[16] However, designers, builders, architects, and engineers have not been satisfied with trying to clone old traditions, nor have they sought to entirely revolutionize the building type, but incorporate the latest science and construction methodologies into modern incarnations of the iconic building type.

A recent development in heavy timber construction is the advent of cross-laminated timber (CLT), which is a flexible building system applicable for many

[14] Anderson-KM Builders, Department of Business Development and Marketing. *Anderson-KM Builders Commences Work on Historic Renovation of Faribault Woolen Mill and Ground Up Multifamily Housing Project. Anderson-KM Builders*, (Minneapolis: 12 Sept. 2011).

[15] Peter J. DeCarlo, 'Nearly 150 Years of Wool Manufacturing in Faribault.' *MinnPost*, 18 February 2014.

[16] Ciro Cuono, 'Heavy Timber Framing.' *Civil + Structural Engineer*, November 2011.

assembly types. It is a prefabricated solid wood panel made of a minimum of three orthogonally bonded layers of solid-sawn or structural composite lumber. The layers are then laminated by gluing the members together with structural adhesives to form a solid rectangular timber, primarily used in roof, floor, and wall assemblies.[17] Glue-laminated wood (glulam) is a composite timber that consists of several smaller pieces of stressed wood laminated together, which is claimed to be stronger than a similar-sized solid timber. Glulam sections are available either in curved or straight shapes and can come in lengths of up to 65 ft, with dimensions of up to 6 in. by 24 in.[18] The most commonly used new timber technology is laminated veneer lumber (LVL), which as the name implies is a composite product composed of multiple layers of wood veneers laminated together. Similar to glulam, LVL has demonstrated increased structural performance over comparable solid timbers.[19]

In terms of construction methods, two new systems have emerged and become rather prominent. The first is post-tensioned timber construction, which harkens back to its descendant, post-and-beam construction. This method, like its predecessor, utilizes large-dimensional post-and-beam timbers but also utilizes steel post-tensioning cables embedded in the timbers to increase the strength, stability, and stiffness of the timber frame, which results in the possibility of greater interior spans. This method usually utilizes glulam or LVL timbers.[20] The more new-age system is known as panelized construction and heavily uses CLT panels as its primary structural members. However, structures built using this method are often restricted to a maximum height of five stories; otherwise, the building requires a concrete core to maintain stability.[21]

To combat the ever-present danger of fire, which may or may not be exacerbated by these new structures' deviations from traditional methods, materials, and occupancies, it has been deemed to be of critical importance to outfit them with the latest in fire protection technology as well. Not seen on most historic heavy timber buildings, in a code-dominated building environment, automatic fire sprinkler systems have become all but mandatory installations. What was also not seen in old heavy timber buildings but has been making appearances in new buildings is the physical manifestation of the encapsulation principle. Encapsulation, either complete or limited, applies layers of gypsum plasterboard over the timber structural members in an attempt to not only protect them from the thermal effects of fire, but prolong their survival under fire conditions.[22]

---

[17] Steven W. Judson, *Board Memo – Building Codes Division*, (Olympia: State of Oregon, 2014), p. 1.

[18] Robert Gerard, David Barber, and Armin Wolski, *Fire Safety Challenges of Tall Wood Buildings*, (Quincy: Fire Protection Research Foundation, 2013), p. 24.

[19] Gerard, pp. 25–26.

[20] Gerard, pp. 30–33.

[21] Gerard, pp. 33–37.

[22] Andrew Buchanan, Birgit Ostman, and Andrea Frangi, *Fire Resistance of Timber Structures*, (Canterbury: University of Canterbury, 2014), pp. 9–10.

Aside from innovations in technology and attractive aesthetics, builders have expressed a renewed interest in constructing new heavy timber buildings for their proven performance under extreme stress, be it from fire, heavy loads, or volatile weather conditions.[23] Additionally, with the popularization of composite timbers, it has become both more cost effective and environmentally sustainable to build in heavy timber.[24] It has been found that it is cheaper for builders to erect a low-high-rise building, one that ranges in height from 75 to 125 ft, to build in heavy timber than in steel and concrete. It has also been determined that harvesting and manufacturing a renewable resource such as timber is far more environmentally friendly and reduces our carbon footprint.[25] What is more is that the age-old enemy of heavy timber, the labor-intensive construction process, is minimized with the employment of current timber technologies and is on par with other prefabricated construction methods, even heavy timber's old successor, steel and concrete.[26]

A varied group of existent modern heavy timber structures litter the American landscape. However, few of these new builds resemble the traditional monolithic heavy timber building. They incorporate a wide array of new shapes and a massive variety of sizes, but almost exclusively depart from the utilitarian identity of heavy timber, instead attempting to provide aesthetic value. These structures encompass buildings of gargantuan proportions such as the 220,000-ft^2 Branson Convention Center in Missouri and Oakland's Cathedral of Christ the Light (Fig. 7.5) which frames its 1500-seat auditorium with 26 110-ft glulam ribs. Yet they also encompass the small and quaint such as the 1240-ft^2 Saint Joseph Adoration Chapel at Belmont Abbey in North Carolina.[27]

As impressive as some of these structures are, timber-oriented designers have also set their sights on possible heavy timber structures that are far more grandiose. Notably, Minnesota-born architect, Benton Johnson, of Chicago's Skidmore, Owings & Merrill, proposed a design for a 395-ft tall, 42-story heavy timber skyscraper.[28] However, it is important to note that many of these new builds are not true heavy timber buildings in the strictest sense. They often require the use of concrete foundations and even steel beams.[29]

---

[23] WoodWorks, *Innovations in Timber Construction*, (Washington: Wood Products Council, 2011), p. 3.

[24] David Barber, and Robert Gerard, 'High-Rise Timber Buildings.' *Fire Protection Engineering*, 1 July 2014.

[25] Barber and Gerard, 2014.

[26] *A Study of Alternative Construction Methods in the Pacific Northwest*, (Seattle: Mahlum Architects, 2014).

[27] WoodWorks, pp. 5–11.

[28] Don Jacobson, 'Heavy Timber to Rise Again in Construction.' *Minneapolis Star Tribune*, 19 September 2013.

[29] Nalina Moses, 'Lumbering into the Sky: Building Tall with Wood.' *AIA*, The American Institute of Architects, (2015).

Fig. 7.5 Interior of Oakland's Cathedral of Christ the Light (Photograph by Carol M. Highsmith's America, Library of Congress, Prints and Photographs Division)

Rescued from obsolescence and returned to life, American heavy timber construction continues on in the modern age. Whether revived altogether, repurposed for a new mission, or constructed from the ground up, a renaissance of this storied construction type has gripped the nation, a veritable return to its structural roots and a well-deserved new beginning for a long-serving building type that had been all but left for dead.

Chapter 8
A Hope for Survival

In the preceding chapters, we have both chronicled the extensive history of American heavy timber construction and examined its exceptionally fascinating design characteristics. We have been witnesses to the demise of the type and its commonplace relegation to obsolescence and decay. We have seen these iconic and dominating structures demolished and reduced to rubble by fire. Conversely, we've seen a movement in which these old workhorse buildings are given a new lease on life through renovation and occupancy changes.

However, for purists, there is still hope for the relatively untouched and uninterrupted survival of this distinctive type of building. In this country, even decades after their widespread abandonment and the adoption of more economically and technologically efficient construction alternatives, pristine and largely unaltered heavy timber buildings survive. They may well be hidden in plain sight as unassuming monuments for an era of construction long since passed. As such, this chapter will take an in-depth look at one unique example of Upper Midwestern heavy timber construction, one that is as elegant and detailed as it is robustly built. Therefore, this chapter will provide a detailed architectural analysis and a detailed history of a single American heavy timber building.

Guardian Angels Church in Chaska, Minnesota, is precisely what one would expect, the city of Chaska's Catholic parish church. The current church was preceded by several smaller churches, each one growing progressively larger. In 1885, the current church was constructed and has served residents of Chaska, with relatively minimal interruptions, for nearly 130 years. Despite its rather advanced age, the church is used extensively in its support of well over 1000 families. On any given week, it facilitates nine Masses, houses a variety of social functions, and familiarizes school children with Catholic Mass procedure. This type of continuous use may well explain why this particular heavy timber structure is so well preserved.

Guardian Angels is located in downtown Chaska, at the corner of 2nd Street West and North Cedar Street. The church is the heart of a reduced yet still bustling Catholic community. While the adjacent friary sits empty and the nunnery has been

J. Heitz, *Fire Resistance in American Heavy Timber Construction*,
DOI 10.1007/978-3-319-32128-8_8

converted into the parish office, Guardian Angels Church still serves as the backbone of a small Catholic complex which contains a school for kindergarten through eighth grade.

Guardian Angels Church is an incredible example of the late-nineteenth-century German-American Gothic Revival architecture. The imposing nine-bay church is approximately 140 ft in overall length with a width of approximately 72 ft when measured across the transepts. Similar to many Catholic churches in the region and perhaps in general, Guardian Angels features a cruciform floor plan (Fig. 8.1).

Aside from being constructed almost exclusively from locally-produced materials, the large church is an exceptional example of American heavy timber construction. Thick bricks combined with framing from solid timbers of the largest order make it a marvel of structural redundancy. Its architect is unknown, although he or she would undoubtedly recognize it even after over a century (Fig. 8.2). Certainly, it has changed in order to conform to modern needs and utilities, even recovering from a devastating fire, yet the church remains in a condition very similar to that of its inception.

The massive church, clad in cream-colored Chaska brick and complete with a 162-ft-tall steeple, commands attention. It is truly an imposing structure that utterly dominates the landscape. Even in a county of 376 mile^2 and at the age of 140, Guardian Angels remains the tallest building in Carver County. So noticeable is Guardian Angels that even though it is nestled into the Minnesota River Valley, its steeple can be easily spotted from miles away (Fig. 8.3).

The current heavy timber structure is the veritable capstone of a parish that had grown exponentially. The Guardian Angels parish found its genesis in 1842 when a French missionary, Father Augustin Ravoux, established roots in the Minnesota territory.[1] His first church, St. Francis Xavier, a small log chapel, was sold to Catholics in Wabasha in 1845 after the Dakota threatened to burn it.[2] In 1858 when Minnesota obtained statehood, Father Bruno Riss and the Benedictines came to Chaska to serve the largely German and Dutch population of the newly incorporated village of Chaska and by 1860 had constructed the first church at the intersection of Third Street and Ash Street.[3] In 1865, it was decided that the small and simple brick building with wooden plank pews should be replaced. By 1870, a much larger 30-ft-by-90-ft rectangular church built out of brick had been completed.[4] As work began on the foundation of the second church, the *Weekly Valley Herald* stated:

> Dimensions of the building will be 30×90 with a basement for school purposes and will cost at least $6000. The building will be of brick, and being situated on one of the most beautiful and commanding sites in town, will when finished, be not only an ornament to the place, but an honor to the Church....It will be the best Church structure in the county when finished.[5]

---

[1] Father Paul Jarvis, *Guardian Angels Catholic Church*, (Chaska: Father Paul Jarvis, 2008), p. 10.

[2] Jarvis, p. 10.

[3] Jarvis, p. 10.

[4] Theresa Fahey, *Guardian Angels Church, Chaska, Minnesota: 150th Anniversary, 1858–2008*, (Chaska: Guardian Angels, 2008), pp. 14–15.

[5] 'New Catholic Church: It Has Been Decided to Build a New Catholic Church in Chaska.' *Weekly Valley Herald* [Chaska], 11 September 1868.

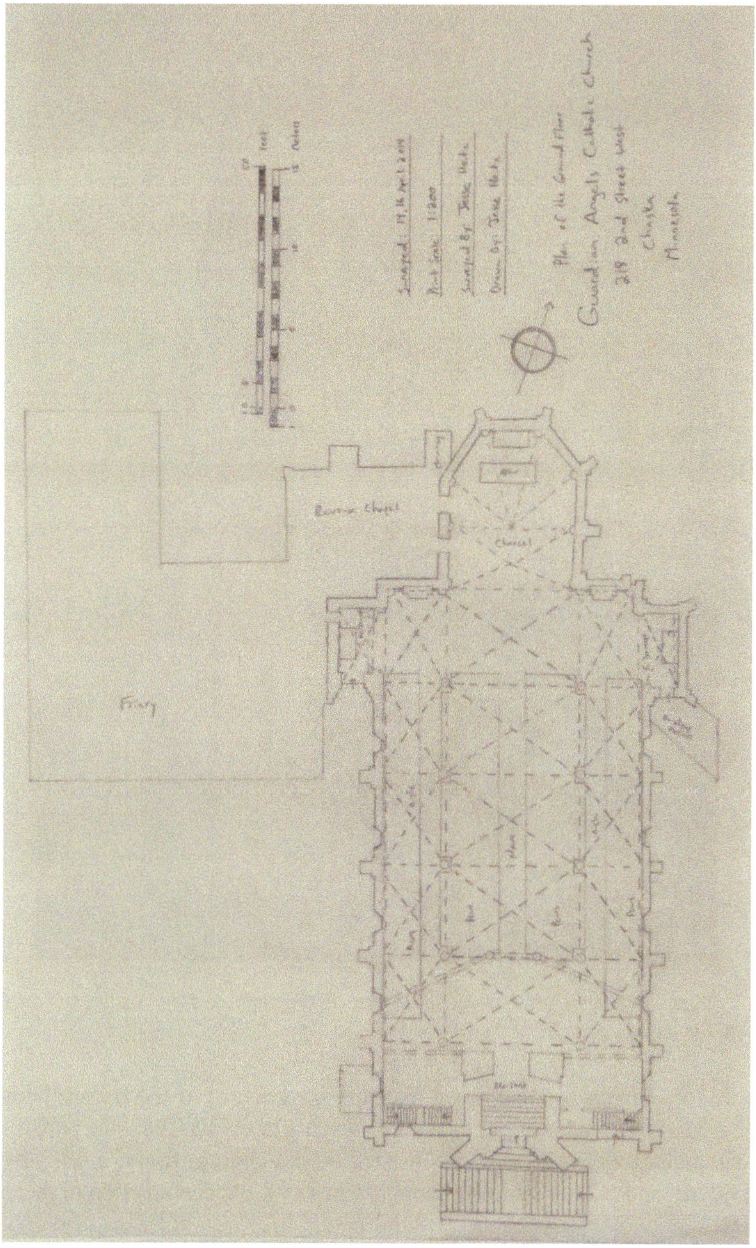

Fig. 8.1 A sketch of the Guardian Angels floor plan, measured and drawn by the author

Fig. 8.2 Guardian Angels prior to the 1902 fire (Photograph courtesy of the Carver County Historical Society)

By 1876, Franciscans had taken over the governance of the parish, building a 55-ft-by-33-ft friary adjacent to the church 4 years later.[6] By 1882, the parish served approximately 160 families[7] and consisted of the church, friary, a 60-ft-by-40-ft schoolhouse, and a 28-ft-by-28-ft nunnery.[8] In 1884, the construction of the current

[6] Edward D. Neill, and Charles S. Bryant, *History of the Minnesota Valley: Including the Explorers and Pioneers of Minnesota*, (Minneapolis: North Star Pub., 1882), p. 359.

[7] Neill, p. 359.

[8] Neill, p. 359.

Fig. 8.3 Guardian Angels' steeple viewed from 2 miles south at dusk (Photograph taken by the author)

church began on the plot occupied by the 1870 church.[9] Although it is interesting to note that Guardian Angels, both the second and the 1885 churches, violated Catholic architectural precedent by building along a north-south axis as opposed to the traditional west-east axis. The *Weekly Valley Herald* reported on the plans for the new church (Fig. 8.4), stating:

> Our new Catholic Church, under the charge of Father Clementius, will be one of the finest structures in this part of the country…. The church will be in the shape of a cross, and will be 142 ft. long by 52 ft. wide and 72 ft. wide inside the cross. The church will be 67 ft. high from the foundation to peak of roof, or 39 ft. from foundation to eaves. The steeple will be 162 ft. high and will contain the best chime of bells to be found in this part of the country….[10]

By October of 1885, the new church had been completed, in the process cannibalizing the bricks of the 1870 church.[11] So large was the new church that its western transept butted up against the western wall of the friary. From a purely external view, one can notice irregularities that serve as physical evidence of the friary

[9] 'New Buildings: Will Begin Working on the New Church.' *Weekly Valley Herald* [Chaska], 2 April 1885.

[10] 'The New Catholic Church.' *Weekly Valley Herald* [Chaska], 24 April 1884.

[11] Jarvis, p. 10.

Fig. 8.4 The church under construction (Photograph courtesy of the Carver County Historical Society)

predating the 1885 church. The western transept features no prominent buttress, much less the diagonal buttresses featured on the stand-alone eastern transept. One also notices that the wall of the western transept features no evidence of a window ever having existed, counter to the eastern transept. Also the walls of the western transept can be clearly seen overhanging the roof of the original friary building. Additionally, when one explores the friary attic, there is a window looking directly into the wall of the western transept, at such proximity that one could open the window and touch the transept (Fig. 8.5).

In 1899, an unnamed fresco artist was brought up from St. Louis, Missouri, to paint the interior of the church.[12] Unfortunately, any photographs or written accounts of the frescoes' appearance have not been uncovered. As such, one cannot speculate as to what they looked like.

As stated previously, the current church was erected in 1885, arguably in the wake of the apex of German immigration into the United States. By this time, immigration into the United States from Northwestern Europe, notably Germany, was soundly eclipsed by a wave of immigration from Southern, Central, and Eastern Europe, perhaps most notably Italy, the Austro-Hungarian Empire, and the Russian Empire.[13] What is interesting is that, although entirely expected, each of these

[12] Fahey, p. 30.

[13] John Powell, *Encyclopedia of North American Immigration*, (New York: Facts On File, 2005), p. 302.

Fig. 8.5 Junction of the western church wall, transept, and friary (Photograph taken by the author)

distinct ethnic groups brought with them their own unique structural culture, which in turn was routinely demonstrated, and Guardian Angels is no exception.

While one of the oldest Catholic churches of its class, it shares a striking resemblance with several other area Catholic churches built by German-Americans. Churches like Freeport's Church of the Sacred Heart (1904) possesses a similar vaulting scheme, buttress shape and formation, and exterior wall adornment.[14] Pierz's St. Joseph's Church (1886–1888) possesses a similar look, but is composed of darker brick.[15] St. Michael's Catholic Church of the town of the same name (1890–1891), another German-American Gothic Revival church, features a similar appearance and is even constructed out of Chaska brick, although it is edged with dressed stone.[16] All of these churches break sharply with the Catholic churches of neighboring communities constructed by Central and Eastern European-derived people, which are largely of a Baroque style, although often much more reserved than those found in Europe.

Upon first glance and when viewing Guardian Angels through merely a superficial exterior, the church is by no means extravagantly adorned, festooned with elaborate decoration (Fig. 8.6). Some might argue that it is plain; however, that is a somewhat overly simplistic view. While it is splendid in its inherent external simplicity, the structure is far more impressive when one examines its inner workings.

---

[14] Alan K. Lathrop, and Bob Firth, *Churches of Minnesota: An Illustrated Guide* (Minneapolis: University of Minnesota Press, 2003), p. 55.

[15] Lathrop, p. 145.

[16] Lathrop, p. 163.

Fig. 8.6 Guardian Angels' western wall and north face of the original friary (Photograph taken by the author)

Guardian Angels Church is a rather unique heavy timber structure for two very poignant reasons. First, it is because of the incredible dedication and outright impressive vision required by the small community to construct such a large heavy timber structure. Second, the building is the physical embodiment of Chaska, making use of the town's natural resources and infrastructure.

The church is finely constructed out of the high-quality cream-colored Chaska bricks, an industry which produced 3.5 million bricks per day at its zenith in 1902.[17] The bricks are arranged into American bond, with seven courses of stretchers between a single course of headers. The bricks themselves do well to date the church. Many are weathered, with coarse faces and rounded edges. Some are stained and cracked. The mortar is faded in color and less than uniform in its appearance (Fig. 8.7).

True to heavy timber construction form, it is framed with locally felled hardwood timbers, as Chaska sits in the state's "Big Woods" region, a roughly 100-mile-by-40-mile strip of dense forest.[18] As such, Guardian Angels is a wonderful example of heavy timber construction, as we have discussed, a popular construction type for

[17] Carver County Historical Society, *Farmhouses in Carver County*, (Waconia: Carver County Historical Society, 2011), pp. 3–4.

[18] Minnesota Department of Natural Resources, *Big Woods: Subsection Profile*, (St. Paul: Minnesota Department of Natural Resources, 2006). http://files.dnr.state.mn.us/assistance/nrplanning/bigpicture/cwcs/profiles/big_woods.pdf. (accessed 1 April 2014)

Fig. 8.7 Eastern wall of the church near the junction with the eastern transept. The American bond pattern of the brick and its varying condition is evident (Photograph taken by the author)

large buildings at high risk of fire damage.[19] The timbers are only visible when one crawls into the attic or up the steeple. Additionally, when one sees that both the attic and steeple are large plenums, it calls into question the church's adherence to the basic tenets of heavy timber construction's modern conception. This in and of itself instills an impression regarding age that such construction almost certainly predates any sort of fire code governing heavy timber, which certainly existed in full by 1915.[20] However, when examining the roof and steeple platform decking thickness, as well as the thickness of the lath and plaster creating the vaults, one could make a convincing argument that this structure satisfies heavy timber's standards for separations.

These massive structural timbers (Fig. 8.8), with posts, beams, wall plates, and even principal rafters measuring roughly 8 in. by 11 in., are indicative of age alone as it would be exceptionally difficult to obtain a solid hardwood timber of such dimensions, with laminated timbers being virtually the only option in recent decades. Another telling detail exhibited by the timbers which may indicate age is that critical structural timbers, such as king posts, are notched in order to accommodate the principal rafters, something that would not be acceptable under current structural engineering conventions (Fig. 8.9).

---

[19] Brannigan, p. 233.

[20] C. G. Smith, *Building Code Recommended by the National Board of Fire Underwriters*, (New York: National Board of Fire Underwriters, 1915), p. 3.

Fig. 8.8 Heavy timbers in the roof assembly above the eastern aisle, flanked by vaults (Photograph taken by the author)

Fig. 8.9 Close-up of a notched king post in the church attic (Photograph taken by the author)

Fig. 8.10 Iron strap and bolts affixed to a post which frames the steeple (Photograph taken by the author)

The condition of the timbers themselves may provide clues to their age. The principal structural timbers visible in the attic are quite dark with some crudely cut and shaped and some cracking. Those found in the steeple show demonstrable evidence of weathering. Moreover, the metal components connecting structural timbers to timbers and to the thick brick walls also date the church and help to solidify its identity as a heavy timber structure. The straps and bolts connecting structural timbers appear to be hand-forged wrought iron, often exhibiting significant surface rust (Fig. 8.10), something one would expect to see replaced by steel as the twentieth century wore on.

The same is the case with the tie rods that hold the load-bearing walls to the timber frame, although the tie rods protruding from the wall plate on each side of the buttresses (Figure 8.11) are a key identifier of heavy timber construction and are threaded at the end yet feature no accompanying nut or anchor plate (Fig. 8.12), unlike the friary. Furthermore, there is no evidence of rust stains on the surrounding brick; however, such stains may very well have been removed during the sandblasting of the brick in the 1990s (Fig. 8.13).

The wood itself, given its tight rings and striations, mixed with its reddish color, indicates that it may well be the heartwood of red oak. Such a conclusion is further supported by the fact that the Minnesota "Big Woods" area was composed of nearly 10 % red oak at the time of the church's construction and subsequent reconstruction,[21]

[21] Minnesota Department of Natural Resources, (2006).

Fig. 8.11 Tie rods exiting the eastern wall of the church alongside the buttresses (Photograph taken by the author)

Fig. 8.12 Close-up of the tip of a tie rod along the eastern wall of the church (Photograph taken by the author)

Fig. 8.13 Anchor plate and rust staining on the friary; both are absent features on the church itself (Photograph taken by the author)

and that such a species of hardwood has been a staple of heavy timber construction.[22] The darkening of these timbers also indicates advanced age.[23]

On 7 October 1902, fire erupted from within the church. The fire left only the thick brick exoskeleton of the church standing. The spire fell off and crashed into the neighboring Du Toit house.[24] While some argued that the fire was a result of arson, a more viable theory finds that the fire was caused by a malfunction of the recently installed furnace.[25] The loss was estimated at $50,000 dollars, and the church was rebuilt in a year at a cost of $34,000 dollars.[26]

Interestingly enough, the evidence of charred timbers and brick in the basement room which houses the power plant shows that fire-damaged structural fabric was reused and built upon (Fig. 8.14). This unusual find highlights the robust and fire-resistant nature of heavy timber. Similarly, inside the steeple, there are numerous

[22] Jack Sobon, *Build a Classic Timber-Framed House: Planning and Design, Traditional Materials, Affordable Methods*, (Pownal: Storey Communications, 1994), p. 352.

[23] Adelbert P. Mills, Lloyd F. Rader, and Harrison W. Hayward, *Materials of Construction: Their Manufacture and Properties*, (New York: Wiley, 1939), p. 623.

[24] 'Monstrous Fire Demon Totally Destroys Chaska's Beautiful Guardian Angel Church.' *Weekly Valley Herald* [Chaska], 9 October 1902.

[25] 'Two Contracts Let: Court House & Catholic Church to Have Heating Plants.' *Weekly Valley Herald* [Chaska], 17 July 1902.

[26] Fahey, p. 30.

Fig. 8.14 Charred timber and brick in the mechanical room adjacent to the church (Photograph taken by the author)

glazed bricks. This could very well have been their condition upon installation, but their concentration might indicate that they survived the fire.

By late 1902, work to precisely rebuild the church was progressing.[27] Certainly, there were slight deviations from the original plan. For example, it had been decided to enlarge the basement and to pour a concrete floor.[28] Interestingly when one looks at the foundation of the original friary, it is crudely composed of essentially rubble and mortar (Fig. 8.15). Therefore, one might be able to infer that the foundation of the original 1885 church was similar. However, there are no written accounts to support such a conclusion.

Furthermore, to reduce the risk of fire and improve the aesthetics of the church, a small one-story rectangular structure housing a small chapel and a new below-ground power plant and chimney[29] was joined to the church at the western side of the chancel and to the northern wall of the friary (Fig. 8.16); however, these alterations do not appear to have compromised the integrity of the church's form.

---

[27] 'Work on the Church: Will Have the Basement Ready, to Have Christmas Services in the Basement.' *Weekly Valley Herald* [Chaska], 20 November 1902.

[28] 'The Corrogated Ceiling Is Being Put on in the Basement of the Church.' *Weekly Valley Herald* [Chaska], 15 January 1903.

[29] 'Improvements During the Year: Lists Buildings Erected or Improved in Chaska.' *Weekly Valley Herald* [Chaska], 11 February 1904.

Fig. 8.15 Rubble and mortar foundation found in the mechanical room adjacent to the church (Photograph taken by the author)

Fig. 8.16 Chimney and Ravoux Chapel butted up against the chancel, nave, western transept, and noticeably shorter eastern wing of the friary (Photograph taken by the author)

Fig. 8.17 Passageways into the Ravoux Chapel from the church chancel (Photograph taken by the author)

This addition is particularly vexing. When standing in church's chancel, a wide doorway had been awkwardly cut in the chancel's western wall, allowing for direct passage into the Ravoux Chapel (Fig. 8.17). An additional doorway appears to have been installed a little further down the wall which leads visitors past another doorway into the Ravoux Chapel and into a kitchen facility. This little structure in essence creates a passageway from the church and into the friary. Internally, it seems to be of lightweight construction inconsistent with both the church and friary and has been heavily modernized. Even Guardian Angels hasn't been able to escape modernization completely. To be sure, further additions and alterations would follow.

Butted up against this structure and the church is a tall chimney that ostensibly pipes into the mechanical room below which houses the power plant, replacing a

Fig. 8.18 Hatch on the power plant's chimney (Photograph taken by the author)

smaller chimney that was attached to the western wall of the church just before the transept. The American bond pattern of the brick lines up perfectly between the chimney and the chapel building, but the rough joining of the chapel's cornice to the chimney, and the near obstruction of one of the chapel's windows by the chimney, clearly indicates that they are two distinct structures, but that the chapel most likely preceded the chimney. Additionally, an iron hatch on the chimney featuring heavy rusting (Fig. 8.18), combined with both structures' brick possessing virtually identical coloration and wear, supports the notion that the chimney was built immediately after the chapel.

By January of 1903, the basement had been finished and had been outfitted with pews to conduct Mass.[30] By October of 1903, the church was once again operational with the addition of a spacious choir loft in the shape of an irregular hexagon and pipe organ. In order to support the fore-end of the choir loft, two nondescript cast-iron columns were installed a little over 6 ft apart, effectively providing a gateway down the central aisle in the nave. During this time, the church was also wired for electricity.[31]

For nearly two decades, the heavy timber church remained unaltered from its reconstructed form, even in the face of potential disaster. In 1925, the church survived yet another catastrophe when a tornado struck downtown Chaska. It appears that the schoolhouse sustained the worst of the damage, having its southwest corner demolished. In 1947, the bell tower was struck by lightning which blew out the

[30] Fahey, p. 32.

[31] 'George W. Henk, the Electrician Has Been Preparing the Electric Light Chandeliers for the Catholic Church.' *Weekly Valley Herald* [Chaska], 12 February 1903.

motors controlling the bells, yet the clocks still registered the correct time, and no other damage was reported.[32] Several years later in the 1950s, the church's spacious basement was overhauled and converted into a recreational area for students.[33] Somewhere in this period, the original stairway to the church entrance was rebuilt, shifting from a direct linear brick stairway into the current form that empties into a small landing then proceeds horizontally to the ground. In 1978, the 75-year-old pipe organ was overhauled, and electronic control was installed.[34]

It wasn't until the dawn of the 1990s that this monolith incurred grievous injury. In 1991, maintenance efforts were undertaken on the church. Its clock faces were refinished and lightning rods were added, the cross sitting atop the steeple was seal coated, the walls were resurfaced which included ill-advised sandblasting of the brick, and the whole of the church was re-roofed. The interior of the church was also repainted.[35]

The steel sheeting on the roof was surely replaced and is well documented, and it also appears that the steel coverings on the buttress offsets were also replaced at the same time. The offsets feature similar amounts of rust and are attached through nails tacked into what appears to be caulking (Fig. 8.19). At this time, it could be that the finials atop the forward facing buttresses were removed and simply capped

Fig. 8.19 An offset on one of the church's buttresses (Photograph taken by the author)

[32] 'Silenced by Lightening: Tower Clock Jolted, the Tower Clock Was Struck by Lightning.' *Weekly Valley Herald* [Chaska], 12 June 1947.

[33] 'New Parish Hall Gym (Photo).' *Weekly Valley Herald* [Chaska], 18 October 1956.

[34] Jarvis, p. 13.

[35] 'Of New Roofs, Lighted Clock Faces & Fundraisers.' *Chaska Herald*, 15 August 1991.

Fig. 8.20 Decking, common rafters, and purlin in the church roof (Photograph taken by the author)

with steel. Likewise, the crosses resting atop both transepts' gables, and the one atop the chancel, were removed altogether.

While in the attic of the church, one can cleanly view the undercarriage of the roof. The purlins are most certainly postfire original. However, the decking and common rafters are more perplexing. They are not as weathered, stained, and warped as those found in the original portion of the friary. Yet the fact that the rafters are true 2-in.-by-6-in. timbers indicates they may very well predate the 1991 re-roofing (Fig. 8.20). This discrepancy may also be a matter of upkeep. The friary has been vacant for nearly two decades, where the church has been in constant use, ensuring that it receives prompt maintenance.

While a true admirer of heavy timber construction can appreciate the beauty of this church from a purely exterior and structural view, the full breadth of its magnificence can be witnessed on the interior, where art truly meets brawn. Many of these carefully crafted accoutrements appear to be original, which adds to the authenticity of this structure. The large recessed double-paned stained glass sash windows with trefoils encased in brick jams with wooden interior sills, appear to be relatively unchanged (Fig. 8.21). The rib vaults appear to be fundamentally unchanged as well (Figs. 8.22, 8.23, and 8.24).

While the diagonal ribs in all of the vaults have a corrugated sectional appearance (Fig. 8.25), the lath and plaster and framing appear to be original when viewed in the attic. Further evidence is seen when one looks up at the surface of both the vaults and walls and a light wave-like appearance from buckling lath can be observed (Fig. 8.26). Given the appropriate and even spacing of the lath when viewed from the attic and absence of cracking plaster, one is inclined to believe that the buckling is due to age (Fig. 8.27).

Fig. 8.21 Window along the western aisle (Photograph taken by the author)

Six years later in 1997, church leadership, with the often-misunderstood architectural demands of Vatican II in mind[36], considered renewing the church while attempting to maintain its historic integrity. In 2000, an extensive renovation program was instituted which included installing air conditioning, re-carpeting, new lighting, new liturgical art, replacing the pews, and repainting.[37]

There is indeed photographic evidence from the early twentieth century that indicates that many of the interior features have remained the same (Figs. 8.28 and 8.29). The basic wooden parapets with simple inlaid arcading on the choir loft and western transept gallery are the same, as is the paneling that covers the beams running directly below them. The large hollow wooden interior piers with Corinthian capitals appear to be original as well. However, with timber posts visible in the attic above and corresponding cast-iron columns in the basement below, it is unclear which of those materials composes the core of the piers. Furthermore, the interior paint scheme has been largely kept the same, save the removal of a fresco above the transverse rib in front of the chancel and change of color on the vaults in the chancel (Figs. 8.30 and 8.31).

[36] Michael S. Rose, 'Did Vatican II Promote Reform of Church Architecture?", (Dellachiesa.com, 2004).

[37] Jarvis, p. 13.

Fig. 8.22 Vaulting along the central space looking southward to the choir loft (Photograph taken by the author)

Fig. 8.23 Vaulting springing from a column in the nave (Photograph taken by the author)

Aside from aesthetic value, the existence of these fittings helps to support the argument that continuous use has greatly aided in the preservation of this wonderful specimen. Physical elements alone provide much evidence that many features are post-fire originals. For example, the wooden steps leading from the western transept

Fig. 8.24 Vaulting along the western aisle in the nave facing southward (Photograph taken by the author)

into the friary (Fig. 8.32) and the wooden steps on both sides of the south wall leading up into the choir loft feature excessive wear, something that can only be explained by heavy use over a significant period of time. Similarly, fixtures such as door knobs on the doors leading to both of the aforementioned places feature incredible wear (Fig. 8.33). Additionally, wear and slight warping of the exposed hardwood floors below the pews indicates that they were not replaced when the church floors were re-carpeted.

One can also clearly see that several features were the result of the aforementioned turn-of-the-century renovation program. New lighting was installed in the bosses of the vaults as well as tucked away in the abacus of each pier. On the exterior of the building, a handicap-accessible ramp and overhang was added to the south

Fig. 8.25 Corrugated appearance of the vault ribs (Photograph taken by the author)

face of the eastern transept, and the excellent condition of the concrete and lack of corrosion on the steel posts lead one to believe that such an addition belongs to this particular time period (Fig. 8.34). While the original fabric of this heavy timber structure has been altered, its march toward modernity has been quite gradual.

Despite the inherently robust nature of heavy timber construction, it is not immune to damage. Thankfully and unlike so many heavy timber buildings around the nation, Guardian Angels' state of disrepair was addressed in the first decade of the twenty-first century. In 2008, maintenance efforts were undertaken in earnest stop the decay of the church's brick walls. Such a terrific level of deterioration had been brought about by the 1991 sandblasting episode, which, in addition to destroying the mortar between the bricks, actually removed the tough external crust of the

Fig. 8.26 Wavy appearance of the lath and plaster walls in front of the chancel (Photograph taken by the author)

Fig. 8.27 Laths composing a transverse rib as seen from the attic (Photograph taken by the author)

bricks themselves, leaving a softer portion of the brick exposed to Minnesota's notoriously fierce weather conditions.[38]

Until 1995, the Franciscans were around and took it upon themselves to make repairs.

However, it has been noted that their often quick and efficient repairs were long-term hazards. For example, they used Portland cement to tuck-point the brick walls instead of traditional lime mortar. The cement's inherent rigidity did not allow for the expansion and contraction of the brick caused by the routine 120° Fahrenheit temperature shifts that Minnesota sees throughout the course of a given year. The end result was extensive damage to the joints, creating voids between bricks large

[38] Mark Olson, 'Tricky Bricks: Guardian Angels Crafts Plan for Crumbling Façade.' *Chaska Herald*, 11 August 2008.

Fig. 8.28 Undated early-twentieth century photo of the church's central space facing southward (Photograph courtesy of the Carver County Historical Society)

enough to stick an adult man's fingers in. Foliage was even found to be growing out of the parapets (Fig. 8.35).[39]

Work was immediately undertaken to restore the Chaska brick walls, utilizing the City of Chaska and the parish's existing stores of Chaska brick, before moisture and the resultant rotting ravaged the interior of the building. In October of 2009, local mason, Patrick Sieben, began replacing damaged bricks and repairing crumbling mortar. As he delved into the work, it became apparent that such a task would be far more involved than originally anticipated.

Sieben's work lasted well into 2013, replacing approximately 2000 bricks and applying 15,000 lb of mortar in the steeple alone (Fig. 8.36). The bricks used, all authentic Chaska bricks, were recovered from several demolished chimneys in local

[39] Mark Olson, 'Tricky Bricks: Guardian Angels Crafts Plan for Crumbling Façade.' *Chaska Herald*, 11 August 2008.

Fig. 8.29 Central space of the church facing southward as seen in 2014 (Photograph taken by the author)

Fig. 8.30 Undated photograph of the church's central space facing the altar (Photograph courtesy of the Carver County Historical Society)

Fig. 8.31 Vaulting in the chancel and the *top* of the altar in 2014 (Photograph taken by the author)

Fig. 8.32 Steps leading from the western transept into the friary (Photograph taken by the author)

homes.[40] When one examines the exterior brick surfaces, it is easy to spot replacement bricks, which are not weathered and have crisper lines and smooth mortar joints.

Currently, Guardian Angels Catholic Church has been faithfully restored largely to its 1902 post-fire form maintaining its historic fabric and integrity, with certain

[40] Mark Olson, 'Work Continues on Downtown's Most Visible Landmark.' *Chaska Herald*, 7 June 2013.

Fig. 8.33 Heavy wear is evident on the door knob which is affixed to a door to the stairwell to the choir loft (Photograph taken by the author)

Fig. 8.34 Handicap-accessible entrance attached to the eastern transept (Photograph taken by the author)

Fig. 8.35 Brick in disrepair along the western wall (Photograph taken by the author)

modernizations to facilitate twenty-first century conveniences. So dedicated were these repair and renovation efforts that they arguably saved the old church from being closed down and a new structure built.[41] It still serves its bustling Catholic community and still serves as one of Carver County's most notable landmarks. Perhaps more importantly, it stands as one of the premier surviving examples of American heavy timber construction, albeit serving an occupancy type far different than the industrial occupancy the type is commonly associated with.

[41] Heron M. Estrada, 'Saving Chaska's Guardian Angels.' *Minneapolis Star Tribune*, 2 March 2010.

Fig. 8.36 The church steeple, with brick in a wide range of conditions (Photograph taken by the author)

Chapter 9
Conclusion

Heavy timber construction occupies a unique place in American building history. They are a building type that is truly American. It is a construction type that physically embodied the nation from which it was spawned. During a tumultuous time in American history, where industrial progress was in jeopardy of being tamed by fire, a brilliantly innovative design that was heavy timber rose from the ashes. During this period, the construction type echoed its client, American industry, and served as a transition from the small and marked by the artistic expression of craftsmen to the large and utilitarian mass-produced built by fleets of laborers. Heavy timber was streamlined and standardized, establishing itself in every corner of the United States, effectively being ingrained in local culture.

These monolithic structures rose to heights and sizes unseen in American industry, providing the space and infrastructure to propel the United States from a small, largely rural, and uncoordinated economy to a status as one of the world's foremost economic powers. These buildings bore the wounds of the American economy's identity change, one that took it away from manufacturing. Many were left to suffer inevitable dilapidation and destruction. However, some were ultimately saved and brought back from the verge of extinction. What is undeniable is that heavy timber construction served its purpose well. These structures served their intended purpose by providing both durability and proven fire resistance and along the way became a staple of the American built environment (Fig. 9.1).

J. Heitz, *Fire Resistance in American Heavy Timber Construction*,
DOI 10.1007/978-3-319-32128-8_9

Fig. 9.1 Minneapolis' Pillsbury Complex photographed in the twenty-first century (Photograph by Carol M. Highsmith's America, Library of Congress, Prints and Photographs Division)

Bibliography

Ahlgren, Carol. 2008. *Staff Report—Ceresota Building*. Minneapolis: Minneapolis Heritage Preservation Commission.

Anderson-KM Builders, Department of Business Development and Marketing. 2011. *Anderson-KM Builders Commences Work on Historic Renovation of Faribault Woolen Mill and Ground Up Multifamily Housing Project. Anderson-KM Builders*. Minneapolis, September 12.

Andreas, A.T. 1884. *History of Chicago: From the Earliest Period to the Present Time*. Chicago: A.T. Andreas.

Anfinson, Scott F. 2003. Archaeology at the Riverfront: Unearthing the Invisible. *Minnesota Historical Society* 320–331.

Ayers Dean, Edward L. 1992. *The Promise of the New South: Life after Reconstruction*. New York: Oxford University Press.

Barber, David, and Robert Gerard. 2014. High-Rise Timber Buildings. *Fire Protection Engineering*, July 1.

Bivens, Matthew. 2004. *National Register of Historic Places Registration Form: Crunden-Martin Manufacturing Company*. St. Louis: Landmarks Association of St. Louis.

Born, George Walter. 2006. *Preserving Paradise: The Architectural Heritage and History of the Florida Keys*. Charleston, SC: History.

Braddee, Richard W. 2000. *Six Career Fire Fighters Killed in Cold-Storage and Warehouse Building Fire-Massachusetts*. Washington: National Institute for Occupational Safety and Health.

Bradley, Betsy H. 1999. *The Works: The Industrial Architecture of the United States*. New York: Oxford University Press.

Brannigan, Francis L., and Glenn P. Corbett. 2007. *Brannigan's Building Construction for the Fire Service*, 4th ed. Sudbury, MA: National Fire Protection Association.

Brown, Claudia R., and M. Ruth Little. 2000. *National Register of Historic Places Nomination Form—American Tobacco Company Manufacturing Plant*. Raleigh: North Carolina State Historic Preservation Office.

Bryant, Tim. 2013. As Cupples 7 Goes Down, Cupples 9 Thrives. *St. Louis Post-Dispatch*, July 9.

Buchanan, Andrew, Birgit Ostman, and Andrea Frangi. 2014. *Fire Resistance of Timber Structures*. Canterbury: University of Canterbury.

1837. The Builder's Manual. *Mechanics' Magazine, and Journal of the Mechanics' Institute*, p. 198.

Byers, Christine. 2011. St. Louis Blaze That Burned 18 Hours Now Out, Under Investigation. *St. Louis Post-Dispatch*, December 9.

Carver County Historical Society. 2011. *Farmhouses in Carver County*. Waconia: Carver County Historical Society.

J. Heitz, *Fire Resistance in American Heavy Timber Construction*,
DOI 10.1007/978-3-319-32128-8

1871. The Chicago Calamity. *The New York Times*, October 21, p. 11.
1871. Chicago: Curious Effects of the Heat of the Conflagration. *The New York Times*, November 17.
1871. Chicago: High Rates of Insurance—Bad Faith of Insurance Companies. *The New York Times*, November 11, p. 4.
1871. Chicago: Progress in Clearing Away the Rubbish at the Burnt District. *The New York Times*, November 8, p. 6.
1871. Chicago: The Progress of Reconstruction. *The New York Times*, November 23, p. 2.
Ching, Frank. 2007. *Architecture: Form, Space, and Order*. Hoboken, NJ: Wiley.
1902. The Church Work Progressing Nicely. *Weekly Valley Herald* [Chaska], December 18.
City of Chicago. 1984. *Landmark Designation Report*. Chicago: Commission on Chicago Historical and Architectural Landmarks.
1871. The City of Chicago: Slow Progress of Rebuilding. *The New York Times*, December 8, p. 2.
Clement, Winston Wallace. 2014. *Standardization in the Lumber Industry: Trade Journals, Builder's Guides and the American Home*. Philadelphia: University of Pennsylvania.
Coffey, Mike. 2011. Building Construction: How It Effects Your Firefighting. *Carolina Fire Journal*.
Commonwealth vs. Thomas S. Levesque, Superior Court of Massachusetts, March 2002.
Condit, Carl W. 1966. *The Chicago School of Architecture: A History of Commercial and Public Building in the Chicago Area, 1875–1925*. Chicago: University of Chicago.
Connors, Edward. 2012. *National Register of Historic Places Registration Form—Heaton & Cowing Mill*. Riverside: Edward Connors and Associates.
1903. The Corrogated Ceiling Is Being Put on in the Basement of the Church. *Weekly Valley Herald* [Chaska], January 15.
2014. Cultivation of a Tobacco Empire. *North Carolina Historic Sites*, North Carolina Department of Cultural Resources Office of Archives & History.
Cuono, Ciro. 2011. Heavy Timber Framing. *Civil + Structural Engineer*, November.
DeCarlo, Peter J. 2014. Nearly 150 Years of Wool Manufacturing in Faribault. *Minn Post*, February 18.
1921. Department of Architectural Engineering: Posts and Lintels. *The American Architect and the Architectural Review* 120: 205.
Dunn, Vincent. 2007. *The Strategy of Firefighting*. Tulsa, OK: Pen Well.
Environmental Protection Agency. 2003. *Former Civil War Textile Mill Is Given New Economic Life with the Help $775,000 in Funding from the EPA*. December 3.
Estrada, Heron M. 2010. Saving Chaska's Guardian Angels. *Minneapolis Star Tribune*, March 2.
1879. Explosions of Flour Dust. *New York Evangelist*, January 23, p. 7.
FJA. 1876. Checking the Fire Fiend. *Scientific American* XXXIV(17): 261.
Fahey, Theresa. 2008. *Guardian Angels Church, Chaska, Minnesota: 150th Anniversary, 1858–2008*. Chaska: Guardian Angels.
Father Paul Jarvis. 2008. *Guardian Angels Catholic Church*. Chaska: Father Paul Jarvis.
Fazio, Michael W., Marian Moffett, Lawrence Wodehouse, and Marian Moffett. 2008. *A World History of Architecture*. Boston: McGraw-Hill.
1921. Fighting Fire. *Outlook*, October 5, p. 163.
1871. The Fire Fiend. *The New York Times*, October 8, p. 5.
1871. Fire-Proof Buildings—Views of the Sculptor, Powers. *The New York Times*, December 30, p. 4.
1854. Fire-Proof Buildings. *Circular* 3(116): 462.
1871. Fire-Proof Cities. *Every Saturday: A Journal of Choice Reading*, November 4, p. 434.
1857. Fire-Proof Structures. *Flag of Our Union* 12(26): 205.
1999. *Fire-Rated Systems*. Tacoma: APA—The Engineered Wood Association.
Fiske, Henry A. 1907. *Quarterly of the National Fire Protection Association* 1: 74–75.
Fitch, James Marston. 1966. *The American Building and the Environmental Forces That Shape It*, 2nd ed. Boston: Houghton Mifflin.

Fossum, Paul R. 1930. Early Milling in the Cannon River Valley. *Minnesota Historical Society*, 271–282. Northfield: Carleton College.

Frame, Robert M., III. 1978. Mills, Machines, and Millers. *Minnesota Sources for Flour-Milling Research*, 152–162. Minneapolis: Minnesota History.

Francisco, Mollee. 2015. How Sweet It Is. *Chaska Herald*, April 20.

Franckling, Ken. 1987. Historic New England Buildings Increasingly Fill Today's Needs: Reuse of Old Mills, Factories Has Become Economic Phenomenon in New England. *Los Angeles Times*, December 8.

Friedman, Donald. 1995. *Historical Building Construction: Design, Materials, and Technology*. New York: W.W. Norton.

Garvin, James L. 2001. *A Building History of Northern New England*. Hanover: University of New England.

Gebhard, David, and Tom Martinson. 1977. *A Guide to the Architecture of Minnesota*. Minneapolis: University of Minnesota.

1903. George W. Henk, the Electrician Has Been Preparing the Electric Light Chandeliers for the Catholic Church. *Weekly Valley Herald* [Chaska], February 12.

Gerard, Robert, David Barber, and Armin Wolski. 2013. *Fire Safety Challenges of Tall Wood Buildings*. Quincy: Fire Protection Research Foundation.

Gilham, Paul C. 2015. *Fire-Resistive Design of Exposed Timber Structures*. Tualatin: Western Wood Structures.

Gladhill, Bethany. 2006. *Chaska Historic Context Study*. Chaska: Thomas R. Zahn & Associates.

1871. Great Fire at Paterson. *The New York Times*, December 14.

Greenwood, Richard E. 2011. Providence One Hundred Years Ago—The Industrial Heyday. *City of Providence City Archives*.

Gruson, Lindsey. 1983. Illegally Stored Propane Is Blamed for Blast That Killed 6 in Buffalo. *The New York Times*, December 29.

Handlin, David P. 1985. *American Architecture*. London: Thames and Hudson.

Harris, Gale. 2003. *Brown Building (Originally Asch Building)*. New York: Landmarks Preservation Commission.

History of Fire and Fire Protection in Nineteenth Century Woonsocket. *City of Woonsocket*, Woonsocket Fire Department.

Hitchcock, Lawrence. 1915. *A Treatise on Interior Plastering, for the Use of Architects, Owners, Contractors*. Pittsburgh: Jackson-Remlinger.

Hodgson, Fred T. 1909. *Light and Heavy Timber Framing Made Easy; Balloon Framing, Mixed Framing, Heavy Timber Framing, Houses, Factories, Bridges, Barns, Rinks, Timber-roofs, and All Other Kinds of Timber Buildings*. Chicago: F.J. Drake.

International Association of Fire Chiefs, National. 2015. *Fire Inspector: Principles and Practice*. Sudbury: Jones & Bartlett Learning.

Jacobson, Don. 2013. Heavy Timber to Rise Again in Construction. *Minneapolis Star Tribune*, September 19.

Judson, Steven W. 2014. *Board Memo—Building Codes Division*. Olympia: State of Oregon.

Kimbell, R. G. 1924. Architectural Engineering: Building Code Limitations on Floor Areas in Buildings. *The American Architect and the Architectural Review* 125: 301.

Kleiner, Fred S., and Christin J. Mamiya. 2004. *Gardner's Art Through the Ages: A Concise History*. Belmont, CA: Wadsworth.

Langenbach, Randolph. 2010. *Better than Steel? (Part 2): Tall Wooden Factories and the Invention of "Slow-burning" Heavy Timber Construction*. Oakland: Conservationtech Consulting.

Lang-Kummer, Karen, and A. Rebecca Harrison. 1982. *National Register of Historic Places Inventory-Nomination Form—Shockoe Valley and Tobacco Row Historic District*. Richmond: Virginia Historic Landmarks Commission.

1879. The Largest Flour Mill in the World. *Scientific American* XLI(19): 291.

Lathrop, Alan K., and Bob Firth. 2003. *Churches of Minnesota: An Illustrated Guide*. Minneapolis: University of Minnesota Press.

Law, Ben, and Lloyd Kahn. 2010. *Roundwood Timber Framing: Building Naturally Using Local Resources*. East Meon, Hampshire: Permanent Publications.

1879. The Law of Dust Explosions. *Scientific American* XLI(18): 282.

Ledermann, Robert P. 2011. *State Street: One Brick at a Time*. Charleston: History Press.

Lescaze, William. 1937. The Meaning of Modern Architecture. *The North American Review* 244: 110.

Lewis, Michael J. 2002. *The Gothic Revival*. New York, NY: Thames & Hudson.

Lockley, Neil, and Jo Marceau. 2000. *The Story of Architecture*. New York: Dorling Kindersley.

Longswisch, Cynthia. 1985. *National Register of Historic Places Inventory-Nomination Form: Cupples Warehouse District*. St. Louis: Landmarks Association of St. Louis.

1871. Mayor Medill: His Inaugural Message to the New Council of Chicago. *The New York Times*, December 8, p. 2.

McGinnis, F. S. 1928. Use and Conservation of Our Timber Supply. *Overland Monthly and Out West Magazine*, November.

McMurry, Sally Ann, and Annmarie Adams. 2000. *People, Power, Places*. Knoxville: University of Tennessee.

McNamara, Denis R., and James Morris. 2005. *Heavenly City: The Architectural Tradition of Catholic Chicago*. Chicago: Liturgy Training Publications.

Meyer, David R. 2003. The Roots of American Industrialization, 1790–1860. *Economic History Services*.

1981. *MHC Reconnaissance Survey Town Report Brockton*. Boston: Massachusetts Historical Commission.

1986. Mill District Development Begins. *Preservation Matters* 2, p. 1.

Miller, Lee H. 1929. Towering Steel. *Youth's Companion*, May.

Mills, Adelbert P., Lloyd F. Rader, and Harrison W. Hayward. 1939. *Materials of Construction: Their Manufacture and Properties*. New York: Wiley.

Minnesota Department of Natural Resources. 2006. *Big Woods: Subsection Profile*. St. Paul: Minnesota Department of Natural Resources. http://files.dnr.state.mn.us/assistance/nrplanning/bigpicture/cwcs/profiles/big_woods.pdf. Accessed 1 Apr 2014.

1902. Monstrous Fire Demon Totally Destroys Chaska's Beautiful Guardian Angel Church. *Weekly Valley Herald* [Chaska], October 9.

Mittal, Manju. 2013. Explosion Hazard and Safety in Industries Handling Grain Products. *Journal of Engineering Research and Studies* IV(3).

1839. Monmouth Mutual Fire Insurance Company. *Maine Farmer and Journal of the Useful Arts* 7(51): 403.

Moses, Nalina. 2015. Lumbering into the Sky: Building Tall with Wood. *AIA*, The American Institute of Architects.

National Park Service. *National Register of Historic Places*. Washington, D.C.: U.S. Department of the Interior. http://www.cr.nps.gov/nr/. Accessed 1 Apr 2014.

Naum, Christopher J. 2015. *National Fire Fighter Near-Miss Reporting System May-Structural Collapse*. Washington: Command Institute.

Naum, Christopher. 2015. Lessons from the Fireground-1958: Wooster Street Fire & Collapse. *Fireground Leadership*, February 16.

Neill, Edward D., and Charles S. Bryant. 1882. *History of the Minnesota Valley: Including the Explorers and Pioneers of Minnesota*. Minneapolis: North Star Publication.

1885. New Buildings: Will Begin Working on the New Church. *Weekly Valley Herald* [Chaska], April 2.

1868. New Catholic Church: It Has Been Decided to Build a New Catholic Church in Chaska. *Weekly Valley Herald* [Chaska], September 11.

1991. Of New Roofs, Lighted Clock Faces & Fundraisers. *Chaska Herald*, August 15.

Norman, John. 2005. *Fire Officer's Handbook of Tactics*. Tulsa, OK: Penn Well.

Olson, Mark. 2010. Preserving a Chaska Landmark: Guardian Angels Restoring Church. *Chaska Herald*, February 11.

Olson, Mark. 2008. Tricky Bricks: Guardian Angels Crafts Plan for Crumbling Façade. *Chaska Herald*, August 11.

Olson, Mark. 2013. Work Continues on Downtown's Most Visible Landmark. *Chaska Herald*, 7 June 7.

Owens, L.L. 2008. *The Great Chicago Fire*. Edina: ABDO Publications.

Patterson, James. 1993. *Simplified Design for Building Fire Safety*. New York: Wiley.

Paul, C.E. 1916. *Heavy Timber Mill Construction Buildings*. Chicago: Engineering Bureau, National Lumber Manufacturers Associations.

Peckham, S. F. 1878. On the Explosion of the Flouring Mills in Minneapolis. *American Journal of Science and Arts* 16(94): 301.

Pennefeather, Shannon M. 2003. *Mill City: A Visual History of the Minneapolis Mill District*. St. Paul, MN: Minnesota Historical Society.

Pereira, Rui M., and William J. Hausman. 2014. *Railroads and Economic Growth in the Antebellum United States*. Williamsburg: College of William and Mary.

Pessen, Edward. 1980. How Different from Each Other Were the Antebellum North and South? *The American Historical Review* 85(5): 1119–1149.

Petersen, Penny A., and Marjorie Pearson. 2007. *Architecture and Historic Preservation on the Minneapolis Riverfront*. Minneapolis: Hess, Rose.

Pistor, Nicholas J.C. 2013. Cupples Complex Shrinks as St. Louis Struggles to Keep Historic Buildings Standing. *St. Louis Post-Dispatch*, July 9.

Powell, John. 2005. *Encyclopedia of North American Immigration*. New York: Facts On File.

2006. *Preserving America's Heritage: Companion to National Historic Preservation Act*. Washington: Advisory Council on Historic Preservation.

Principles of Building Construction: Combustible, Tech. no. FEMA/USAF/NFA-PBC:C-SM, 2nd ed. National Fire Academy.

1875. The Proposed Building Law. *Chicago Tribune*, March 11.

Ramsdell, H. J. 1871. Fireproof Materials. *Scientific American* XXV(25): 385.

Randall, Frank A. 1999. *History of the Development of Building Construction in Chicago*. Champaign: University of Illinois.

Rauschart, Lisa. 2012. Cover Story: Old Buildings "Repurposed" as Unique Housing. *The Washington Times*, September 6.

Rayfield, Jo Ann. 1997. Tragedy in the Chicago Fire and Triumph in the Architectural Response. *Northern Illinois University Library*. Northern Illinois University.

1871. Rebuilding Chicago. *The New York Times*, November 29, p. 2.

1922. Reinforced Concrete Caps for Wooden Posts. *The American Architect and the Architectural Review* 121: 236.

Rifkind, Carole. 1980. *A Field Guide to American Architecture*. New York: New American Library.

1872. The Risk of Buildings from Fire. *Scientific American* XXVII(11): 165.

Ristau, Toni. 1975. *Mill Architecture in Paterson, NJ: A Culmination of the Empirical Tradition in Construction*. Proc. of Symposium on Industrial Archaeology, New Jersey, Paterson, 59–67. Paterson: Northeast Historical Archaeology.

RLGA Technical Services. 2006. *Building Classification—Part 2: Construction Types*. Arizona: Ronald Geren.

Roberts, Nicholas W. 2004. *Building Type Basics for Places of Worship*. Hoboken, NJ: Wiley.

Rose, Michael S. 2004. Did Vatican II Promote Reform of Church Architecture? Dellachiesa.com.

http://www.foscc.com/downloads/mr/Did%20Vatican%20II%20Promote%20the%20Reform%20of%20Church%20Architecture.pdf. Accessed 13 Apr 2014.

Routley, J. Gordon. 1987. *Detroit Warehouse Fire Claims Three Firefighters*. United States Fire Administration: Emmitsburg.

Rumbarger, Janet, Richard Vitullo, and Charles George Ramsey. 2003. *Architectural Graphic Standards for Residential Construction*. New York: Wiley.

Schaffer, Brian. 2008. *Heritage Preservation Commission Staff Report—Washburn A Mill, Mill City Museum*. Minneapolis: Minneapolis Heritage Preservation Commission.

Schwieterman, Joseph P., Dana M. Caspall, and Jane Heron. 2006. *The Politics of Place: A History of Zoning in Chicago*. Chicago: Lake Claremont.

1871. The Secret of Fire-Proof Construction. *The New York Times*, November 26, p. 4.

Sennott, R. Stephen. 2004. *Encyclopedia of Twentieth Century Architecture*. New York: Fitzroy Dearborn.

1947. Silenced by Lightening: Tower Clock Jolted, the Tower Clock Was Struck by Lightning. *Weekly Valley Herald* [Chaska], June 12.

Sinkevitch, Alice, and Laurie McGovern Petersen. 1993. *AIA Guide to Chicago*. New York: Harcourt Brace.

Smith, C. G. 1915. *Building Code Recommended by the National Board of Fire Underwriters*. New York: National Board of Fire Underwriters.

Smith, Michael L. 2000. Heavy Timber Construction: What Firefighters Need to Know. *Firehouse Magazine*, July 1.

Sobon, Jack. 1994. *Build a Classic Timber-Framed House: Planning and Design, Traditional Materials, Affordable Methods*. Pownal: Storey Communications.

Sokoloff, Kenneth L. 1992. Invention, Innovation, and Manufacturing Productivity Growth in the Antebellum Northeast. In *American Economic Growth and Standards of Living before the Civil War*. Chicago: University of Chicago.

1992. Solid Geometry: The Pickwick Mill. *Preservation Matters* 8: 1, 7.

State of Minnesota, Department of Public Safety. 2015. *Construction Types—Definitions*. St. Paul: State of Minnesota.

State of New Jersey, Division of Codes and Standards. 2006. *Types of Building Construction—International Building Code*, 77–78. Trenton: State of New Jersey.

1871. The State of the Manufacturing Trades. *Scientific American* XXV(13): 200.

2014. *A Study of Alternative Construction Methods in the Pacific Northwest*. Seattle: Mahlum Architects.

2003. *Superior Fire Resistance*. Centennial: American Institute of Timber Construction.

Technical Services Division. 1961. *Heavy Timber Construction Details*. Washington, D.C.: National Lumber Manufacturers Association.

Technical Services Information Bureau. 2008. *IBC Building Types*. Orange: Technical Services Information Bureau.

1878. Terrible Explosion. *New England Farmer, and Horticultural Register*, May 11, p. 2.

1868. Test of Fire-Proof Flooring. *Railway Times*, December 5.

2013. Thurston County Washington. *Fire Flow Requirements 2013*. Olympia: Thurston County.

1902. Two Contracts Let: Court House & Catholic Church to Have Heating Plants. *Weekly Valley Herald* [Chaska], July 17.

United States Fire Administration, National Fire Academy. 2013. *Building Construction: Heavy Timber Construction (Type IV)*. Emmitsburg: National Fire Academy.

United States National Park Service. 2014. *Preservation Brief 21: Repairing Historic Flat Plaster Walls and Ceilings National Parks Service*. Washington, D.C.: U.S. Department of the Interior. http://www.nps.gov/tps/how-to-preserve/briefs/21-flat-plaster.htm. Accessed 10 April 2014.

Ver Planck, Christopher. 2009. *District Record—New Wholesale District*. San Francisco: Department of Parks and Recreation.

1871. The Very Latest. *The New York Times*, October 10, p. 1.

Waldheim, Charles, and Katerina Rüedi. 2005. *Chicago Architecture: Histories, Revisions, Alternatives*. Chicago: University of Chicago.

Warner, Gene. 2013. 30 Years Later, Survivors Recount Horror of Propane Blast. *Buffalo News*, December 21.

Warren, S. 1924. Texture in Portland Cement Stucco: Part III. *The American Architect and the Architectural Review* 126: 319.

Watts, Alison. 2000. The Technology That Launched a City. In *Minnesota Historical Society*, 86–97. Minneapolis: Minnesota History.

Weinberg, Meyer. Rise of the Capitalist Class, 1790–1865. *New History*.

White, Robert H. 2013. Post-Fire Analysis of Solid-Sawn Heavy Timber Beams. *Structure Magazine*, November.

White, Robert H. 1985. *Reporting of Fire Incidents in Heavy Timber Structures*. Madison: Forest Products Laboratory.

Willis Group Holdings—North America Strategic Outcomes Practice. 2010. *Managing Risks in Vacant and Idle Buildings*. Chicago: Willis.

Wood Works. 2011. *Innovations in Timber Construction*. Washington: Wood Products Council.

1902. Work on the Church: Will Have the Basement Ready, to Have Christmas Services in the Basement. *Weekly Valley Herald* [Chaska], November 20.

Index

J. Heitz, *Fire Resistance in American Heavy Timber Construction*,
DOI 10.1007/978-3-319-32128-8

MIX
Papier aus verantwortungsvollen Quellen
Paper from responsible sources
FSC® C105338

If you have any concerns about our products,
you can contact us on
ProductSafety@springernature.com

In case Publisher is established outside the EU,
the EU authorized representative is:
Springer Nature Customer Service Center GmbH
Europaplatz 3, 69115 Heidelberg, Germany

Printed by Libri Plureos GmbH
in Hamburg, Germany